Physical Realizations of

Quantum Computing

PHYSICAL REALIZATIONS OF QUANTUM COMPUTING

Are the DiVincenzo Criteria Fulfilled in 2004?
(With CD-Rom)

Osaka, Japan 7 – 8 May 2004

editors

Mikio Nakahara
Shigeru Kanemitsu
Kinki University, Japan

Martti M Salomaa
Helsinki University of Technology, Finland

Shin Takagi
Fuji-Tokoha University, Japan

World Scientific

NEW JERSEY • LONDON • SINGAPORE • BEIJING • SHANGHAI • HONG KONG • TAIPEI • CHENNAI

Published by

World Scientific Publishing Co. Pte. Ltd.

5 Toh Tuck Link, Singapore 596224

USA office: 27 Warren Street, Suite 401-402, Hackensack, NJ 07601

UK office: 57 Shelton Street, Covent Garden, London WC2H 9HE

British Library Cataloguing-in-Publication Data
A catalogue record for this book is available from the British Library.

PHYSICAL REALIZATIONS OF QUANTUM COMPUTING (With CD-ROM)
Are the DiVincenzo Criteria Fulfilled in 2004?

ISBN 981-256-473-X

Printed in Singapore by World Scientific Printers (S) Pte Ltd

To the memory of Martti M. Salomaa
(6 October 1946 – 9 December 2004)

PREFACE

This volume contains papers by invited speakers of the symposium "Quantum Computation: Are the DiVincenzo Criteria Fulfilled in 2004?" held at Kinki University in Osaka, Japan, during the period of May 7-8, 2004. The aim of this symposium was to examine the potential of various physical realizations of a quantum computer in view of the DiVincenzo criteria by experts in each realization.

It is our sad duty to annuonce that Martti M. Salomaa, one of the editors and the contributors, has passed away. We would like to dedicate this volume to his memory. We have reproduced the obituary for Martti Salomaa[a] prepared by his longtime friend Professor Rainer Salomaa and his former student Dr Saku Lehtonen.

Dr David DiVincenzo kindly contributed to this symposium using a video conference system, a prerecorded video tape and a powerpoint file from Chicago. The attaced CD-ROM contains his video contribution and the powerpoint file. Also included in the CD-ROM is the powerpoint file prepared by Akio Hosoya for his contribution to the symposium.

Authors were asked to prepare their contribution in a self-contained way and therefore we expect that each contribution serves as an excellent textbook for one quarter/semester postgraduate course.

We gratefully acknowledge financial support from Kinki University. We would like to thank Professor Megumu Munakata, Vice-Rector of Kinki University and Professor Nobuki Kawashima, Director of School of Interdisciplinary Studies of Science and Engineering, Kinki University, for their interest and support. We would like to thank Ms Zhang Ji for excellent editorial work. One of the editors (MN) would like to thank Yukitoshi Fujimura and Shin-ichiro Ogawa for TeXnical help and Shin-ichiro Ogawa for preparing the movie file in the attached CD-ROM.

Osaka, November 2005

Mikio Nakahara
Shigeru Kenemitsu
Shin Takagi

---

[a]The original obituary is found at http://focus.hut.fi/marttisalomaa3.htm.

Quantum Computation:
Are the DiVincenzo Criteria Fulfilled in 2004?

Kinki Univerisity, Osaka, Japan

---

7 – 8 May 2004

---

7 May

Martti Salomaa (Helsinki Univerisity of Technology, Finland)
DiVincenzo criteria and Beyond

Toshimasa Fujisawa (NTT Basic Research Laboratory, Japan)
Dynamics of single-electron charge and spin in semiconductor quantum dots

Kouichi Semba (NTT Baisc Research Laboratory, Japan)
Superconducting qubits: Experimental forefront and challenges

Frank Wilhelm (Ludwig-Maximilians-Universität, München, Germany)
Superconducting qubits: Theoretical prospects, architectures, and limitations

Hartmut Häffner (Universität Innsbruck, Austria)
Quantum computation with trapped ions - Teleportation with atoms

Akio Hosoya (Tokyo Institute of Technology, Japan)
Complementarity of Entanglement and Interference

8 May

Yasushi Kondo (Kinki University, Japan)
NMR Quantum Computer Experiments at Kinki University

David DiVincenzo (IBM, Thomas J. Watson Research Center, USA)
Prospects for Quantum Computation

Bill Munro (Hewlett-Packard, Bristol, UK)
Quantum Information Processing using single photons

Kae Nemoto (National Institute of Informatics, Japan)
Continuous-variable quantum computation

Frank Wilhelm (Ludwig-Maximilians-Universität, München, Germany)
Alternatives to the standard paradigm - does a superconducting quantum computer have to satisfy the DiVincenzo criteria?

CONTENTS

Preface vii

Obituary: Martti Mikael Salomaa 1

DiVincenzo Criteria and Beyond 3
M. M. Salomaa and M. Nakahara

Single-Electron Charge and Spin Qubit in Semiconductor Quantum Dots 16
T. Fujisawa

Superconducting Quantum Computing: Status and Prospects 38
F. K. Wilhelm and K. Semba

Controlling Three Atomic Qubits 108
H. Häffner, M. Riebe, F. Schmidt-Kaler, W. Hänsel, C. Roos, M. Chwalla, J. Benhelm, T. Körber, G. Lancaster, C. Becher, D.F.V. James and R. Blatt

Liquid-State NMR Quantum Computer: Hamiltonian Formalism and Experiments 127
Y. Kondo, M. Nakahara and S. Tanimura

Optical Quantum Computation 185
K. Nemoto and W. J. Munro

Martti Mikael Salomaa (1946 – 2004)

Professor Martti M. Salomaa (born 6 October 1949), director of the Materials Physics Laboratory at Helsinki University of Technology (TKK), passed away due to an acute fit on the 9th of December 2004.

Salomaa was one of the two Finns receiving, for the first time, a 2-year scholarship from the Finnish Cultural Foundation for studies in the Atlantic College (Wales) in 1966. In addition to an international matriculation exam, the heritage of this period of time for Salomaa includes his passion for boating and the large circle of friends encompassing people from various quarters of international society. One prominent and pleasing assignment for students at the Atlantic College was to design and build a lifeboat achieving a speed higher than 30 knots per hour. At the time, this successful project aroused extensive attention in the British press.

Professor Salomaa graduated from Helsinki University of Technology in 1972 and received his LicTech and PhD degrees in 1974 and 1979, respectively. His dissertation dealt with the theory of superfluidity of ^{3}He in ultralow temperatures. This and his further work was closely connected to the ultralow-temperature research of the Low Temperature Laboratory at TKK. For those efforts, the award of the Körber Foundation (Germany) was granted jointly to Salomaa and his three colleagues to acknowledge their exceptional merits in promoting European science.

Professor Salomaa had an exceptionally extensive international research career. He was unanimously regarded as an appreciated lecturer and an eligible collaborator. For Professor Salomaa, the cumulative time passed as a visiting researcher or a professor in Denmark, Germany, the United States, Switzerland, Austria and Japan extended to several years. The international collaboration resulted in a great number of publications with the other top-ranked researchers.

Salomaa was appointed the professor of theoretical materials physics at TKK in 1994. He rapidly assembled a new group consisting of young active researchers. He had the ability to adopt the newest trends of his discipline. He was also capable of channelling his inexhaustible energy into initiating extremely interesting research projects. As the director of the Materials Physics Laboratory, he immediately transferred these new ideas into the teaching.

In Finland, Salomaa and his theory group were the pioneers in Bose-Einstein condensation and in quantum computing. His group was also among the first national groups studying nanotechnology. In addition to the theoretical research, Salomaa initiated research topics related to modern wireless communications systems. His group studying crystal and surface acoustics grew to be one of the largest in the world. The academic merits of both groups are remarkable.

Professor Salomaa was an inspiring character who succeeded in peopling himself with young talented physicists. He dedicated himself full-heartedly to promoting the careers of his students. In particular, he showed exemplary devotion in his role as an instructor for the various research projects of his students. Professor Salomaa was known for his capability of assuming an immense work load and an uncompromising work attitude. He had an amazing ability to totally immerse himself in matters he considered important, and a willingness to pursue them.

In addition to his work as a university professor, Salomaa's focus in life was his beloved family; his wife Margaretha and his four children, Krista, Maria, Tommi, and Linda. Recently, a family-driven and successful summer cottage building project served for him as a counterbalance for the exhausting office work.

Solving new and intriguing problems, both practical and theoretical, was an endless source of joy and satisfaction for Professor Salomaa. He never got tired with research and creative efforts of any kind.

DiVincenzo Criteria and Beyond*

Martti M. Salomaa

Materials Physics Laboratory, POB 2200 (Technical Physics), Helsinki University of Technology, FIN-02015 HUT, Finland
and
Department of Physics, Kinki University Higashi-Osaka, 577-8502, Japan

Mikio Nakahara

Department of Physics, Kinki University Higashi-Osaka, 577-8502, Japan
E-mail: nakahara@math.kindai.ac.jp

Five DiVincenzo criteria and two additional networkability criteria are introduced. Then current status and prospects of the research on physical realization of quantum computing are reviewed following the ARDA QIST roadmap. Finally some proposals for "beyond the DiVincenzo criteria" are outlined.

1. Introduction

Quantum information processing is an emerging discipline in which information is encoded and processed in a quantum-mechanical system.[1] It is expected to solve some of problems that are impossible for current digital computers to execute in a practical time scale. Although small scale quantum information processing, such as quantum key distribution, is already available commercially, physical realizations of a large scale quantum information processor are still beyond the current technology.

Benioff proposed in 1982 that a quantum mechanical Turing machine might be reversible and dissipated no energy.[2] Feynman in the same year published a paper in which he first recognized a quantum system as a useful resource for information processing.[3] His idea has been developed since then by a small group of people including Deutsch.[4] A revolution took

*This contribution was presented by MMS at the symposium. His unexpected early death, however, made it impossible for him to submit the manuscript. MN has prepared this contribution following MMS's presentation and powerpoint file.

place when Shor announced an efficient algorithm for integer factorization.[5] The impact his work brought about was tremendous since the security of internet communication dependes on the belief that factorization of a large interger is practically impossible.

In classical information theory, information is encoded in a bit, which takes values 0 and 1, while in quantum information theory, 0 and 1 are replaced by orthonormal basis vectors $|0\rangle$ and $|1\rangle$ of a two-dimensional complex vector space. Here information is encoded in a qubit (quantum bit) in the form $|\psi\rangle = \alpha|0\rangle + \beta|1\rangle$, where $|\alpha|^2 + |\beta|^2 = 1$. It is estimated that a quantum computer superior to a digital computer today requires at least $10^2 \sim 10^3$ qubits. Although a quantum computer with a small number of qubits, up to 7, is available in several physical systems, construction of a working quantum computer is still a challenging task. DiVincezo proposed necessary conditions that any physical system has to fulfill to be a candidate for a viable quantum computer.[6] In the next section, we outline these conditions as well as two additional criteria for networkability.

2. DiVincenzo Criteria

In an influential article,[6] DiVincenzo, the keynote speaker of this symposium, proposed five criteria that any physical system must satisify to be a viable quantum computer. We summarize the relevant parts of these criteria, which may be helpful in reading following contributions in this volume.

(1) *A scalable physical system with well characterized qubits.*

To begin with, we need a quantum register made of many qubits to store information. The simplest way to realize a qubit physically is to use a two-level quantum system. For example, an electron, a spin 1/2 nucleus and two polarization states (horizontal and vertical) of a single photon may be a qubit. We may equally employ a two-dimensional subspace, such as the ground state and the first excited state, of a multi-dimensional Hilbert space. In the latter case, special care must be taken to avoid leakage of the state to the other part of the Hilbert space. In any case, the two states are identified as the basis vectors, $|0\rangle$ and $|1\rangle$, of the Hilbert space so that a general single qubit state takes the form $|\psi\rangle = \alpha|0\rangle + \beta|1\rangle$, $|\alpha|^2 + |\beta|^2 = 1$. A multi-qubit state is expenaded in terms of tensor products of these basis vectors. Each qubit must be separately addressable. Moreover it should be scalable up to a large number of qubits. The con-

dition of two states may be relaxed to three states (qutrit) or, more generally, d states (qudit).
A system may be made of several different kinds of qubits. Trapped ions, for instance, may employ (1) hyperfine/Zeeman sublevels in the electronic ground state of ions, (2) a ground and excited states of weakly allowed optical transition, and (3) normal mode of ion oscillation, as qubits. A similar scenario is also proposed for Josephson junction qubits.

(2) *The ability to initialize the state of the qubits to a simple fiducial state, such as* $|00\ldots0\rangle$.

In many realizations, initialization may be done simply by cooling. Let ΔE be the difference between energies of the first excited state and the ground state. At low temperatures satisfying $k_B T \ll \Delta E$, the system is in the ground state with a good precision. Alternatively, we may use measurement to project the system into a desired state. In some cases, we observe the system to be in an undesired state on measurement. Then we may transform the system to the desired fiducial state by unitary transformation.
For some realizations, such as liquid state NMR, however, it is impossible to cool the system down to extremely low temperatures. In those cases, we are forced to use a thermally populated state as an initial state. This seemingly difficult problem may be amended by several methods if some computational resources are sacrificed. We then obtain an "effective" pure state, so-called the pseudopure state, which works as an initial state for most purposes.
Continuous fresh supply of qubits in a specified state, such as $|0\rangle$, is also an important requirement for successful quantum error correction.

(3) *Long decoherence times, much longer than the gate operation time.*

Decoherence is probably the hardest obstacle to building a viable quantum computer. Decoherence means many aspects of quantum state degradation due to interaction of the system with the environment and sets the maximum time available for quantum computation. Roughly speaking, this is the time required for a pure state

$$\rho_0 = (\alpha|0\rangle + \beta|1\rangle)(\alpha^*\langle 0| + \beta^*\langle 1|)$$

to "decay" into a mixed state of the form

$$\rho = |\alpha|^2|0\rangle\langle 0| + |\beta|^2|1\rangle\langle 1|.$$

The value of the decoherence time itself is not very important. What matters is the ratio "decoherence time/gate operation time". For some realizations, the decoherence time may be as short as $\sim \mu$ s. This is not necessarily a big problem provided that the gate operation time, detemined by the Rabi oscillation period and the qubit-coupling strength for example, is much shorter than this. If the typical gate operation time is $\sim$ ps, say, the system may execute $10^{12-6} = 10^6$ gate operations before the quantum state decays.
There are several ways to effectively elongate the decoherence time. A closed-loop control method is called quantum error correcting codes (QECC) while an open-loop control method is called noiseless subsystem and decoherence free subspace (DFS). Both of these methods require extra qubits to implement. Time-optimal implementation of a quantum algorithm is regarded as a method to fight against decoherence without any extra resouce.

(4) *A "universal" set of quantum gates.*

Let $H(\gamma(t))$ be the Hamiltonian of an n-qubit system under consideration, where $\gamma(t)$ collectively denotes the control parameters in the Hamiltonian. The time-development operator of the system is

$$U[\gamma(t)] = \mathcal{T} \exp\left[-\frac{i}{\hbar}\int^{T} H(\gamma(t))dt\right] \in \mathrm{U}(2^n),$$

where $\mathcal{T}$ is the time-ordering operator. Our task is to find the set of control parameters $\gamma(t)$ which implements the desired gate U_{gate} as $U[\gamma(t)] = U_{\text{gate}}$. Although this "inverse problem" seems to be demanding to solve, a well known theorem by Barenco *et al* guarantees that any $\mathrm{U}(2^n)$ gate may be decomposed into single-qubit gates $\in \mathrm{U}(2)$ and CNOT gates.[9] Therefore it suffices to find the control sequences to implement U(2) gates and a CNOT gate to construct an arbitrary gate. Note that a general unitary gate in $\mathrm{U}(2^n)$ is written as a product of an $\mathrm{SU}(2^n)$ gate and a physically irrelevant U(1)-phase. Therefore we do not have to worry about the overall phase and it suffices to concentrate on equivalent $SU(2^n)$ gates. This observation is noteworthy since the NMR Hamltonian, for example, is traceless and is able to generate $SU(2^n)$ matrices only. Single-qubit gates are easily implemented if the one-qubit part of the Hamiltonian assumes two of the $\mathfrak{su}(2)$ generators by properly choosing the control parameters. Implementation of a CNOT gate in any realization is considered to be a milestone in this respect. Note however that any two-qubit gate, that is neither a tensor product of two one-qubit gates nor a SWAP gate, works as a component of a universal set of gates.

Quantum circuit implementation requires less steps if multi-qubit gates acting on n (≥ 3) qubits are empolyed as modules.

(5) *A qubit-specific measurement capability.*

The state after an execution of a computation must be measured to extract the result of the computation. Measurement process depends heavily on the physical system under consideration. For most realizations, projective measurements are the primary method to extract the outcome of a computation. In liquid state NMR, in contrast, a projective measurement is impossible and we have to resort to ensemble averaged measurements. This may cause a problem in some cases. Suppose the sytem is in the state $|\psi\rangle = (|00\rangle + |11\rangle)/\sqrt{2}$ for example. The outcome of a projective measurement of $|\psi\rangle$ is $|00\rangle$ with the probability 1/2. The ensemble averaged measurement, on the other hand, yields *both* $|00\rangle$ and $|11\rangle$ with an equal weight. Measurement in general has no 100% efficiency due to decoherence, gate operation error and many more reasons. If this is the case, we have to repeat the same computation many times to achieve reasonably high reliability.

Moreover, we should be able to send and store quantum information to construct a quantum data processing network. This "networkability" requires following two additional criteria to be satisified.

(6) *The ability to interconvert stationary and flying qubits.*

Some realizations are excellent in storing quantum information while long distant transmission of quantum information might require different physical resources. It may happen that some system has a Hamiltonian which is easily controllable and is advantageous in executing quantum algorithms. Compare this with a current digital computer, in which the CPU and the system memory are made of semiconductors while a hard drive is used as a mass storage device. Therefore a working quantum computer may involve several kinds of qubits and we are forced to introduce distributed quantum computing. Interconverting ability is also important in long distant quantum teleportation using quantum repeaters.

(7) *The ability to faithfully transmit flying qubits between specified locations.*

Needless to say, this is an indispensable requirement for quantum communication such as quantum key distribution. This condition is also important in distributed quantum computing mentioned above.

3. Physical Realizations

There are numerous physical systems proposed as a possible candidate for a viable quantum computer. Here is the list of the candidates;

(1) Liquid-state/Solid-state NMR
(2) Trapped ions
(3) Neutral atoms in optical lattice
(4) Cavity QED with atoms
(5) Linear optics
(6) Solid state (spin-based, charge-based)
(7) Josephson junctions (charge, flux, phase)
(8) Electrons on liquid helium surface
(9) Other "unique" realizations.

ARDA QIST roadmap evaluates each of these realizations as outlined in the next section. Subsequent contributions in this volume give detailed accounts on some of these realizations in the light of the DiVincenzo criteria.

4. ARDA QIST Roadmap

Most of the content in this section are excerpts from the online article "A Quantum Information Science and Technology (QIST) Roadmap, Part 1: Quantum Computation"[7] compiled by Advanced Research and Development Activity (ARDA), Los Alamos, USA.[7] This article is updated annualy and readers are strongly recommended to visit this webpage for new progress.

The roadmap was compiled by the members of a Technology Experts Panel to help facilitate the progress of quantum computation (QC) research towards the quantum computer-science era. In kick-off QIST Experts Panel Meeting, held in La Jolla, California, USA in January 2002, the panel members decided that the overall purpose of the roadmap should be to set as a desired future objective for QC

> to develop by 2012 a suite of viable emerging-QC technologies of sufficient complexity to function as quantum computer-science test-beds in which architectural and algorithmic issues can be explored;

The roadmap has a prescriptive role, namely,

> to identify what scientific, technology, skills, organizational, investment, and infrastructure developments will be necessary to achieve the desired goal, while providing options for how to get there;

and a descriptive function

> by capturing the status and likely progress of the field while elucidating the role that each aspect of the field is expected to play toward achieving the desired goal.

The roadmap also identifies gaps and opportunities, and places where strategic investments should be beneficial. This will be helpful for both researchers and research managers.

The panel members introduced the four-level structure in the roadmap, namely, "high-level goal", "mid-level view", "detailed level summaries", and "summary with panel's recommendation".

4.1. *High-Level Goal*

To set a path to realize the desired QC test-bed ear in 2012, the panel provided five- and ten-year technical goals. The five-year (2007) goal is to

- encode a single qubit into the state of a logical qubit formed from several physical qubits,
- perform repetitive error correction of the logical qubit, and
- transfer the state of the logical qubit into the state of another set of physical qubits with high fidelity,

while the ten-year (2012) goal is to

- implement a concatenated quantum error-correcting code.

Meeting the 2007 high-level goal requires the achievement of

- creating deterministic, on-demand quantum entanglement;
- encoding quantum information into a logical qubit;
- extending the lifetime of quantum information; and
- communicating quantum information coherently from one part of a quantum computer to another,

while meeting the 2012 high-level goal requires ~ 50 physical qubits

- to exercise multiple logical qubits through the full range of operations required for fault-tolerant QC in order to perform a simple instance of a relevant quantum algorithm, and
- to approach a natural experimental QC benchmark: the limits of full-scale simulation of a quantum computer by a conventional computer.

The 2012 goal would extend QC into the test-bed regime. It would also enable quantum simulation as originaly envisioned by Feynman.

4.2. *Mid-Level View*

The roadmap also presents a "mid-level view" to allow informed decisions about future directions to be made for researchers. The "mid-level view" segments the field into the different scientific approaches in view of the DiVincenzo criteria and metrics to capture the promise and characterize the progress towards the high-level goals within each approach.

Table 1 summarizes the "promise criteria" according to the mid-level view of the roadmap. The panel used three symbols (letters "G, Y, R" here) to indicate the value the panel assigned to these criteria.

The panel also presents status of each approach with a set of metrics that represent relevant steps on the way to the 2007- and 2012-year goals. Table 2 shows the "development status metrics" in which the panel assigned again three symbols. The numbers $M.$ and $M.N$ in the table refer to the following criteria:

1. Creation of a qubit

   1.1 Demonstrate preparation and readout of both qubit states.

2. Single-qubit operations

   2.1 Demonstrate Rabi flops of a qubit.

Table 1
The Mid-Level Quantum Computation Roadmap: Promise Criteria.
The column number *k* corresponds to the *k*th DiVincenzo crietrion.
(Reproduced with permission from http://qits.lanl.gov/ .)

QC Approach	The DiVincenzo Criteria						
	Quantum Computation					QCNetworkability	
	#1	#2	#3	#4	#5	#6	#7
NMR	C	B	B	A	B	C	C
Trapped Ion	B	A	B	A	A	B	B
Neutral Atom	B	A	B	B	B	B	B
Cavity QED	B	A	B	B	A	B	B
Optical	B	B	A	B	B	B	A
Solid State	B	B	B	B	B	C	C
Superconducting	B	A	B	B	B	C	C
Unique Qubits	This field is so diverse that it is not feasible to label the criteria with "Promise" symbols.						

Legend: A = a potentially viable approach has achieved sufficient proof of principle
B = a potentially viable approach has been proposed, but there has not been sufficient proof of principle
C = no viable approach is known

2.2 Demonstrate decoherence times much longer than the Rabi oscillation period.
2.3 Demonstrate control of both degrees of freedom on the Bloch sphere

3. Two-qubit operations
   3.1 Implement coherent two-qubit quantum logic operations.
   3.2 Produce and characterize the Bell entangled states.
   3.3 Demonstrate decoherence times much longer than two-qubit gate times.
   3.4 Demonstrate quantum state and process tomography for two qubits.
   3.5 Demonstrate a two-qubit decoherence-free subspace (DFS).
   3.6 Demonstrate a two-qubit quantum algorithm.
4. Operations on $3-10$ physical qubits
   4.1 Produce a Greenberger, Horne, and Zeilinger (GHZ) entangled state of three physical qubits.
   4.2 Produce maximally-entangled states of four or more physical qubits.
   4.3 Quantum state and process tomography.
   4.4 Demonstrate DFSs.

Table 2

The Mid-Level QC Roadmap - Development Status Metrics. (Reproduced with permission from http://qits.lanl.gov/ .)

QC Approach	1	1.1	2	2.1	2.2	2.3	3	3.1	3.2	3.3	3.4	3.5	3.6	4	4.1	4.2	4.3	4.4
NMR		S		S	S	S		S	P	P	S	S	S		P	P	S	S
Trapped Ion		S		S	S	S		S*	S*	P	P*	P	P*		N	P	N	N
Neutral Atom		P		S	P	N		N	N	N	P	N	N		N	N	N	N
Cavity QED		P		S	S	P		P	P	P	N	N	N		P	N	N	N
Optical		P		S	S	S		P*	P	P	S	S	S		P	N	N	N
Solid State:																		
Charged or exitonic qubits		S		S	S	N		P*	P*	P*	N	N	N		N	N	N	N
Spin qubits		S		P	P	N		N	N	N	N	N	N		N	N	N	N
Superconducting		S		S	S	P*		P*	N	P*	N	N	N		N	N	N	N

QC Approach	4	4.5	4.6	4.7	4.8	5	5.1	5.2	6	6.1	6.2	6.3	7	7.1	7.2	7.3	7.4	7.5
NMR		S*	S	S	P		N	P		S*	P	N		N	N	N	P	N
Trapped Ion		N	N	N	N		N	N		N	N	N		N	N	N	N	N
Neutral Atom		N	N	N	N		N	N		N	N	N		N	N	N	N	N
Cavity QED		N	N	N	N		N	N		N	N	N		N	N	N	N	N
Optical		P	N	P	N		N	P*		N	N	N		N	N	N	P*	N
Solid State:																		
Charged or exitonic qubits		N	N	N	N		N	N		N	N	N		N	N	N	N	N
Spin qubits		N	N	N	N		N	N		N	N	N		N	N	N	N	N
Superconducting		N	N	N	N		N	N		N	N	N		N	N	N	N	N

Legend: S = sufficient experimental demonstration

P = preliminary experimental demonstration, but further experimental work is required

N = no experimental demonstration and * = a change in the development status between Versions 1.0 and. 2.0

4.5 Demonstrate the transfer of quantum information (e.g., teleportation, entanglement swapping, multiple SWAP operations etc.) between physical qubits.
4.6 Demonstrate quantum error-correcting codes.
4.7 Demonstrate simple quantum algorithms (e.g., Deutsch-Josza).
4.8 Demonstrate quantum logic operations with faulttolerant precision.

5. Operations on one logical qubit
   5.1 Create a single logical qubit and "keep it alive" using repetitive error correction.
   5.2 Demonstrate fault-tolerant quantum control of a single logical qubit.
6. Operations on two logical qubits
   6.1 Implement two-logical-qubit operations.
   6.2 Produce two-logical-qubit Bell states.
   6.3 Demonstrate fault-tolerant two-logical-qubit operations.
7. Operations on $3-10$ logical qubits
   7.1 Produce a GHZ-state of three logical qubits.
   7.2 Produce maximally-entangled states of four or more logical qubits.
   7.3 Demonstrate the transfer of quantum information between logical qubits.
   7.4 Demonstrate simple quantum algorithms (e.g., Deutsch-Josza) with logical qubits.
   7.5 Demonstrate fault-tolerant implementation of simple quantum algorithms with logical qubits.

Detailed-level summaries are given for each physical realization. They are intended to provide a description of each of the experimental approaches and a description of the likely developments over the next decade.

5. Beyond DiVincenzo Criteria

The DiVincenzo criteria are not necessarily the gospel and some conditions can be relaxed. For example, it is possible to replace unitary gates by irreversible non-unitary gates generated by measurements. This idea is already implemented in linear optics quantum computation.[10] An extreme in this approach must be the "one-way quantum computing", where conditional measurements send an initial "cluster state" to the final desired state.[11]

There have also been active discussions concerning sufficiency of the criteria and what comes beyond the DiVincenzo criteria. Here is the list of some proposals:

(1) Implementation of decoherence-free subsystems/subspaces.
(2) Implementation of quantum error correction.
(3) Fault-tolerant quantum computing.
(4) Topologically protected qubits.

Gottesman's article "Requirements and Desiderata for Fault-Tolerant Quantum Computing: Beyond the DiVincenzo Criteria"[12] also discusses requirements for fault-tolerant quantum computing such as

(1) Low gate error rates.
(2) Ability to perform operations in parallel.
(3) A way of remaining in, or returning to, the computational Hilbert space.
(4) A source of fresh initialized qubits during the computation.
(5) Benign error scaling: error rates that do not increase as the computer gets larger, and no large-scale correlated errors.

It also lists "additional desiderata" for a practical quantum computer such as

(1) Ability to perform gates between distant qubits.
(2) Fast and reliable measurement and classical computation.
(3) Little or no error correlation (unless the registers are linked by a gate).
(4) Very low error rates.
(5) High parallelism.
(6) An ample supply of extra qubits.
(7) Even lower error rates.

Many of the above conditions are necessary for quantum error corrections to work reasonably well.

6. Summary

- Solid state and superconducting qubits are expected to be scalable with current lithography technology. However few-qubit operations have been hardly demonstrated with these systems to date.
- In contrast, few-qubit operations are demonstrated on liquid-state NMR and ion traps, although they are probably not scalable.
- Is DiVincenzo criteria classification sufficient? Generalizations "beyond" DiVincenzo may be included using memory, program, ...
- QIST roadmap, based on the DiVincenzo criteria, is extremely valuable for the identification and quantification of progress in this multidisciplinary field.

Acknowledgment

We would like to thank Japan Society for the Promotion of Science (JSPS) for making MMS's stay at Kinki University possible. One of the authors (MN) would like to thank Dr. Richard J. Hughes for allowing us to use materials from ARDA Quantum Information Science and Technolgoy Roadmap[7] and Dr. Daniel Gottesman for allowing us to use his online material[12] and useful suggestions.

References

1. M. A. Nielsen, and I. L. Chuang, *Quantum Computation and Quantum Information,* (Cambridge University Press, Cambridge, 2000).
2. Phys. Rev. Lett. **48**, 1581 (1982).
3. R. P. Feynman, Int. J. Theor. Phys. **21**, 467 (1982).
4. D. Deutsch, Proc. Royal Soc. Lond.: Ser. A **400**, 97 (1985).
5. P. W. Shor, *Proceedings of the 35th Annual Symposium on the Foundation of Computer Science (FOCS '94)*, (IEEE Computer Society Press, Los Alamitos, California, USA, 1994) 124; quant-ph/9508027.
6. D. P. DiVincenzo, Fortschr. Phys. **48**, 771 (2000). See also the attached CD.
7. http://qits.lanl.gov/
8. L. M. K. Vandersypen, M. Steffen, G. Breyta, C. S. Yannonl, M. H. Sherwood, and I. L. Chuang, Nature **414**, 883 (2001).
9. A. Barenco *et. al.* Phys. Rev. A **52**, 3457 (1995).
10. E. Knill, R. Laflamme and G. J. Milburn, Nature **409**, 46 (2001).
11. R. Raussendorf and H. J. Briegel, Phys. Rev. Lett. **86**, 5188 (2001).
12. D. Gottesman, http://www.perimeterinstitute.ca/personal/dgottesman/FTreqs.ppt. See also D. Aharonov and M. Ben-Or, quant-ph/9906129 and J. Preskill, quant-ph/9712048.

Single-Electron Charge and Spin Qubit in Semiconductor Quantum Dots

TOSHIMASA FUJISAWA[1,2,*]

[1] *NTT Basic Research Laboratories, NTT Corporation, 3-1 Morinosato-Wakamiya, Atsugi, 243-0198, Japan*

[2] *Tokyo Institute of Technology, 2-12-1 Ookayama, Meguro-ku, Tokyo 152-8551, Japan*

[*] *E-mail: fujisawa@will.brl.ntt.co.jp*

Semiconductor quantum dots are often regarded as the building blocks of quantum information systems. Single-electron charge states in a semiconductor double quantum dot provides an artificial two-level system (qubit) that can be manipulated by electronic control signals. Full one-qubit operation is demonstrated with a high-speed voltage pulse, in which strength of decoherence is modified by the voltage so as to allow efficient initialization of the qubit. In contrast, single-electron spin in a quantum dot provides a natural two-level system. Because of energy quantization in a quantum dot, electron spin is free from elastic scattering and thus expected to have long relaxation time. We review our recent understanding of single-electron charge and spin in quantum dots, and describe prospects for quantum information applications.

1. Introduction

Quantum information processing promises an efficient computing scheme that can be performed in a *parallel* fashion by *sequentially* manipulating individual qubits and letting two qubits interact.[1,2] Advances in quantum information processing are being achieved through various experiments on natural atoms and molecules, such as pulsed nuclear magnetic resonance of molecules and optically manipulated atomic states,[3–5] and more recently on artificial quantum systems with nanofabrication capability.[6] Recent nanofabrication technology allows us to fabricate artificial atoms (quantum dots) and molecules (coupled quantum dots), in which atomic (molecular) like electronic states can be controlled with external voltages [7–9]. As pointed out by DiVincenzo,[10] quantum computing requires a physical system that holds scalability, initialization scheme, universal set of quantum logic gates, readout scheme, and extremely long relaxation time. Solid state quantum systems are of growing interest, especially for overcoming the scalability issues that often appear in natural atoms. However, the relaxation (decoherence) time could be shortened by the material characteristics in solid state systems. Generally, strong coupling to the environment allows us to manipulate qubits relatively easily, but it degrades the coherency of the system. This difficulty can be overcome by combining different qubits that compensate for their individual weak points or controlling the coupling to the environment dynamically to minimize the decoherence.

Quantum dots (QDs) provide several kinds of qubits in semiconductor materials. An electron charge qubit is an artificial qubit in which all parameters can be controlled by external voltages.[11,12] Since a single electron can be controlled and measured with a high accuracy by means of the Coulomb blockade effect, electron charge qubits may be useful at the interface of quantum and classical information. An electron spin qubit is a natural two-level system, which is expected to have moderately fast operation time and relatively long relaxation time.[13] This is useful as a central processing unitary (CPU) transformer, where quantum algorithms are processed. Nuclear spins have extremely long relaxation time and are useful for quantum memory device. Conversion scheme between single spin qubit and ensemble of nuclear spins in host material has been proposed.[14] Combination of different quantum systems is interesting in obtaining a realistic quantum computation scheme. Semiconductor QD may be a good model system for studying integrated quantum circuits and quantum interfaces among electron charge qubits (as input/output device), electron spin qubits (as central unit), and nuclear spins (as memory device). Various interactions, such as spin-orbit interactions that couple a charge qubit to a spin qubit, and hyperfine interaction between an electron spin qubit and nuclear spins, have been discussed theoretically to this direction, and fundamental characteristics of these interactions have been measured experimentally.

Here, we restrict the discussion to electron charge qubits and spin qubits, including spin-orbit coupling, and describe recent experimental studies on coherent manipulation, dissipation, and decoherence in semiconductor QDs.

2. Charge qubit in a double quantum dot

2.1. *An electron in a double well potential*

An electron in a double-well potential appears quite often in textbooks of quantum mechanics to consider a quantum mechanical tunneling [See Fig. 1(a)]. This simple model can nowadays be realized in a semiconductor double QD (DQD). When the two QDs are well isolated, the system is well

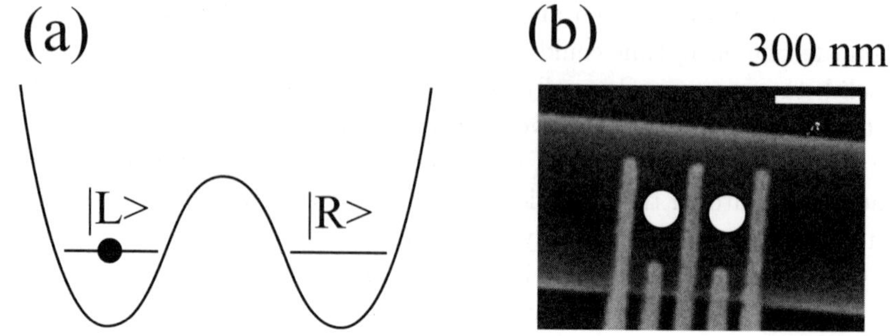

Fig. 1. (a) Schematic of double-well potential. (b) Scanning electron micrograph (SEM) of a double quantum dot device.

described by two charge states, in which an electron occupies one of the states in the left dot ($|L\rangle$) or another in the right dot ($|R\rangle$). If the potential barrier is reasonably prepared to allow quantum mechanical tunneling between the two states, the eigenstates of the system become superpositions of the two states. More generally, a state of an electron in the DQD can be written as a linear superposition of these confined states,

$$|\psi\rangle = \cos\frac{\theta}{2}|L\rangle + \exp(i\phi)\sin\frac{\theta}{2}|R\rangle, \tag{1}$$

where the probability amplitude is described by the coupling angle θ and the phase ϕ. Here, the global phase is removed since it has no physical meaning. This system can be considered a charge qubit in the sense that the basis is an electron charge in the left or the right dot.[12]

A DQD for this purpose can be fabricated routinely using various techniques. High-quality QDs are often fabricated from a two-dimensional electron gas in a GaAs/AlGaAs modulation-doped heterostructure using standard semiconductor fabrication processes, such as electron beam lithography, dry etching, and gate metallization.[15–17] As shown in the scanning electron micrograph of Fig. 1(b), a narrow conductive channel is formed between the upper and lower etched grooves (dark regions). Three tunneling barriers are formed by applying negative voltages to the gate electrodes (the bright vertical lines), leaving the left and right QDs (white circles) between the source and drain electrodes. Optimizing the device structure and dimensions leads to a very simple system where just one electron occupies a double dot.[18] The device we used contains a few tens of electrons basically forming many-body states. In a simplified picture, even if the dots contain more than one electron, one *excess* electron added to the DQD occupies either the right or the left dot. Actually, each charge state involves a ground and excited states corresponding to the orbital degree of freedom in each dot. Therefore, the two-level system (qubit) given by Eq. 1 is a good approximation when only one well-defined state from each dot is considered as $|L\rangle$ or $|R\rangle$. This can be justified under a proper condition where excitation energies, like thermal energy (2.5 μeV at 30 mK for the device we used), are much smaller than the characteristic energies of the dot, i.e., the addition energy (2 - 3 meV), the single-particle excitation energy (50 - 200 μeV), and the electrostatic coupling energy (about 100 μeV).

Figure 2(a) shows a schematic energy diagram of the double-dot device. The energies E_L and E_R are respectively for $|L\rangle$ and $|R\rangle$. The important parameters of the qubit are the energy offset between the confined electron states, $\varepsilon = E_L - E_R$, and tunneling coupling between the two states, T_c. Single electron tunneling current through the DQD is sensitive to these parameters as well as to the tunnel rates, Γ_L and Γ_R respectively for the left and right barriers. Therefore, these parameters can be determined from the current spectrum. Our device geometry is advantageous for independent control of all of these parameters (ε ,T_c ,Γ_L and Γ_R) with five gate electrodes.

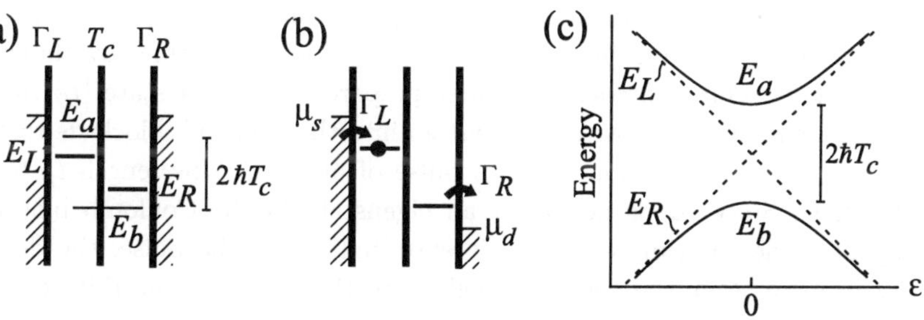

Fig. 2. (a) Schematic energy diagram of the two-level system in a double quantum dot in the Coulomb blockade regime. The bare localized states with energies E_L and E_R are hybridyzed into bonding and antibonding states with energies E_b and E_a . (b) Schematic energy diagram in the transport regime. (c) Schematic energy diagram as a function of energy bias ε.

When the double dot is in the Coulomb blockade region, tunneling processes between it and the electrodes are energetically forbidden. Therefore, to the lowest approximation the coupling to the electrode can be ignored and the simple double-well potential shown in Fig. 1(a) can be applied.

Other decoherence, which will be discussed in Sect. 2.2, is neglected here for simplicity. Then, the Hamiltonian of the system can be written in the simple form

$$H = E_L(t)|L\rangle\langle L| + E_R(t)|R\rangle\langle R| + \hbar T_c\left(|R\rangle\langle L| + |L\rangle\langle R|\right) \tag{2}$$

$$= \frac{1}{2}\varepsilon(t)\sigma_z + \hbar T_c\sigma_x + \text{const.}, \tag{3}$$

where σ_z and σ_x are Pauli matrices in the basis of $|R\rangle$ and $|L\rangle$. When gate voltages are swept so as to cross E_L and E_R as shown by the dashed lines in Fig. 2(c), the eigenstates of the system follows bonding and antibonding states, E_b and E_a respectively, as shown by the solid lines.

In contrast, when a large source-drain voltage is applied, single-electron tunneling current flows through a double dot, and the allowed tunneling processes induce significant decoherence to the charge qubit. The corresponding energy diagram at a large positive voltage is shown in Fig. 2(b). In this case, an electron in the right dot may escape to the right electrode, while another electron may enter the left dot from the left lead. Under the condition $\Gamma_L, \Gamma_R \gg T_c$, the stationary solution of the master equation becomes a density matrix with element $\rho_{LL} \sim 1$, which describes an electron confined in the left dot.[19]

Therefore, the decoherence of the qubit can be switched on (in the transport regime) and off (in the Coulomb blockade regime) by changing the voltage. Our approach is to first prepare a confined state $|L\rangle$ in the transport regime and switch the system into the Coulomb blockade regime by applying a rectangular voltage pulse of an adjustable length t_p to an electrode. Since $|L\rangle$ is no longer an eigenstate in the Coulomb blockade regime, coherent time evolution is expected. After the pulse, the double dot is again set in the transport regime, so that the electron, if it occupies the right dot, contributes to the current. With this scheme, we expect one electron tunneling per pulse at most. By repeating the pulse many times (repetition frequency f_{rep} = 100 MHz), a reasonable pumping current is obtained. Small leakage current that flows through the double dot in the transport regime (during initialization and measurement, but not during state manipulation) is included in the observed current, but this should not degrade the coherency of the qubit.

The coherent dynamics of the charge qubit is determined by the Hamiltonian Eq. 2, which includes two spin matrices. In the Bloch sphere representation of the qubit, the state rotates about the total fictitious magnetic field $B = (2\hbar T_c, 0, \varepsilon)$. In practice, ε can be varied from negative through zero to positive, while $\hbar T_c$ is always positive by definition. Full one-qubit operation can be obtained by adjusting the field to one direction for rotation gate operation and another for phase shift gate operation.

The rotation gate, which changes the occupation of the electron, can be performed by applying B along the x direction ($\varepsilon = 0$), where the state rotates about the x-axis as shown in the inset of Fig. 3(a). The main figure shows the pulse-induced current observed in this situation. A clear oscillation pattern was observed by changing the pulse length, t_p. The oscillation frequency determines $2T_c$, and it can be varied by changing the central gate voltage, as shown in Fig. 3(b). A rectangular pulse that gives a half (quarter) period of the oscillation corresponds to a logical NOT (square-root-NOT) gate for the qubit. The amplitude of the oscillation is approximately half of the ideal case of $ef_{rep} = 16$ pA at the repetition frequency $f_{rep} =$ 100 MHz. [This (small) degradation might have arisen from population of other excited states during initialization. The applied source-drain voltage was even higher than the single particle excitation energy.] The decay of the oscillation is attributed to decoherence, which is discussed in the next subsection.

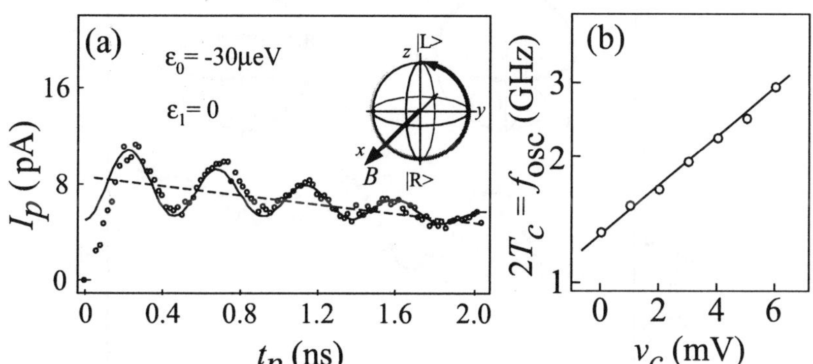

Fig. 3. (a) Current I_p induced by the pulse for the rotation-gate operation. The inset shows the schematic trajectory on the Bloch sphere. (b) Dependence of the coherent oscillation frequency on the central gate voltage.

The ideal phase-shift gate may be realized by applying B along z direction ($\varepsilon \neq 0$ and $T_c = 0$), but this cannot be obtained in experiment. Therefore, we approximate the phase-shift gate by applying $B = (2\hbar T_c, 0, \varepsilon)$ with $\varepsilon \gg \hbar T_c$, where the state precesses more or less about the z-axis. In order to demonstrate the phase-shift operation that does not change electron population, a sharp tipping pulse of length t_t is added at the center of the square pulse for the π-rotation gate (NOT gate).[20] As a result, the first $\pi/2$ pulse prepares the state in a superposition state, $\frac{1}{\sqrt{2}}(|L\rangle - i|R\rangle)$, as shown by the arrow (i) in the inset to Fig. 4. The sharp tipping pulse induces a phase difference ϕ, resulting in a state $\frac{1}{\sqrt{2}}(|L\rangle - i\exp(i\phi)|R\rangle)$, as shown by the arrow (ii). Then, the final state after the second $\pi/2$ pulse becomes $\sin\frac{\phi}{2}|L\rangle + \cos\frac{\phi}{2}|R\rangle$ [See arrow (iii)], whose probability for $|R\rangle$, $\cos^2\frac{\phi}{2}$, should be measured as the corresponding pulse induced current I_p. The phase shift ϕ is approximately given by the area of the pulse as $\phi \simeq eV_t t_t/h$. Actually, we change the height of the tipping pulse V_t while keeping the pulse width constant at $t_t \sim 40$ ps. Figure 4 shows I_p obtained in this way, and the obtained oscillation indicates the phase-shift operation for the qubit. Since the oscillation is obtained with a fixed pulse width, the amplitude decay at higher V_t implies an increased decoherence rate.

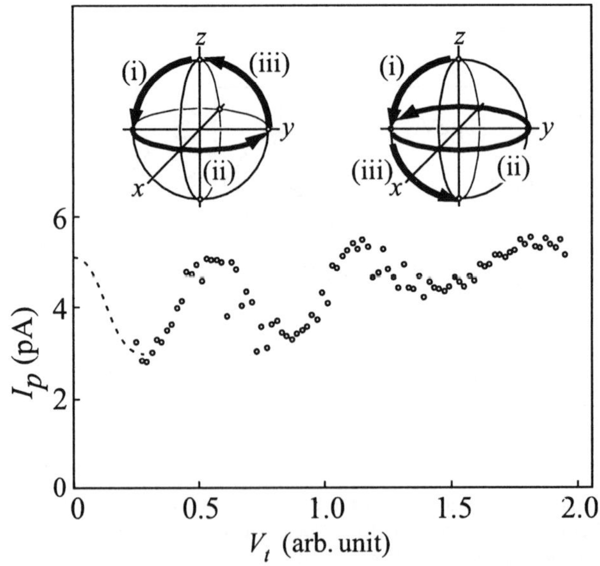

Fig. 4. Dependence of the pulse-induced current I_p on the tipping-pulse height V_t demonstrating the phase-shift operation. The insets show schematic trajectories in the Bloch sphere. The left and right inset corresponds phase shift $\phi = \pi$ and 2π, respectively.

2.2. *Uncontrolled decoherence*

In spite of the successful manipulation of a single-charge qubit, it is actually influenced by uncontrolled decoherence that is present even in the Coulomb blockade regime. Some possible decoherence mechanisms are summarized here.

Firstly, background charge noise ($1/f$ noise) in the sample and electrical noise in the gate voltages cause fluctuation of the qubit parameters ε and T_c, which gives rise to decoherence of the system.[12,21] Low-frequency fluctuation of ε is estimated to be about 1.6 μeV, which is obtained from low-frequency noise in the single-electron current, or 3 μeV, which is estimated from the minimum line width of an elastic current peak at weak coupling limit.[17] Low-frequency fluctuation of $\hbar T_c$ is relatively small and estimated to be about 0.1 μeV at $\hbar T_c = 10$ μeV, assuming local potential fluctuation in the device.[22] Actually, the ε fluctuation explains the decoherence rate observed at the off-resonant condition ($|\varepsilon| \gtrsim \hbar T_c$), which corresponds to the condition for the phase shift gate. The $1/f$ charge fluctuations are usually considered as ensemble of bistable fluctuators, like electron traps, each of which produces a Lorentzian frequency spectrum.[23] The microscopic origin of the charge fluctuators is not well understood, and its magnitude differs from sample to sample, even when samples are fabricated in the same batch. Understanding the fluctuators is a practical and important issue for developing quantum information devices. Recent noise measurement indicates that the fluctuation of ε can be reduced by decreasing temperature as suggested by a simple phenomenological model.[22]

In contrast, the decoherence at the resonant condition ($\varepsilon = 0$), where the rotation gate is performed, is dominated by other mechanisms. Although the first-order tunneling processes are forbidden in the Coulomb blockade regime, higher-order tunneling, namely cotunneling, processes can take place and decohere the system.[24] Actually the cotunneling rate estimated from the tunneling rates is close to the observed decoherence rate, and thus it may be a dominant mechanism in the present experiment.[12] However, since we can reduce the cotunneling effect by making the tunneling barrier less transparent, we should be able to eventually eliminate it in future measurements.

Electron-phonon interaction is an intrinsic decoherence mechanism in semiconductor QDs. Spontaneous emission of an acoustic phonon persists even at zero temperature and causes an inelastic transition between the two states.[17] Actually, the negative background slope shown by the dashed line in Fig. 3(a) corresponds to the inelastic tunneling transition at the off-resonant condition during the initialization/measurement sequence. The energy relaxation time in this case is about 10 ns, but becomes shorter at the resonant condition. We cannot directly estimate the phonon emission rate at the resonant condition from this data, but it may be comparable to the observed decoherence rate. It should be noted that a measurement from a single QD with a slightly different geometry has also shown a phonon emission rate of the order of 10 ns (described in Sect. 3.4).[25] Strong electron-phonon coupling is related to the fact that the corresponding phonon wavelength is comparable to the size of the QD, where the oscillator strength is almost maximized.[17,26] In this sense, electron-phonon coupling may be reduced by using much smaller or much larger QD structures. In addition, polar semiconductors, like GaAs, exhibit a piezoelectric type of electron-phonon coupling, which is significant for low-energy excitations (< 0.1 meV for GaAs).[27] Non-polar semiconductors, like Si, would be preferable for reducing the phonon contribution to the decoherence.

Other mechanisms, such as the electromagnetic environment, have to be considered to fully understand the decoherence. It should be noted that the quality of the coherent oscillation can actually be improved by reducing high-frequency noise from the gate voltages and the coaxial cable. We expect that further studies will exploit ways to reduce some decoherence effects.

2.3. *Charge detection of a DQD*

In contrast to the standard current-voltage measurement covered in the previous section, charge detection measurement directly probes the charge state of QDs. Tunneling barriers between the double dot and the electrodes are not necessary in principle. Actually, single-electron transistors and quantum point contact (QPC) structures have been used as a sensitive electrometer for detecting a charge state in single and DQDs.[18,28–30] In addition, the radio frequency technique provides wide bandwidth for high-speed detection with a high-impedance device.[29,31,32] These techniques are interesting, especially for detecting a charge state in relatively a short time, which permits *single-shot readout* for the charge qubit. Single-shot readout is essential for studying individual states and the correlation of two or more qubit systems.

Figure 5(a) shows a device structure containing a DQD in the upper electrical channel and a QPC charge detector in the lower channel (The dotted regions are the conductive regions). The upper and lower conductive regions are well isolated ($\gg 10$ GΩ) by applying sufficiently large voltage

V_{iso} to the isolation gate, while other neighboring regions are separated by tunneling barriers with finite rates. The DQD in the upper channel shows characteristics similar to that used in observing coherent charge oscillations. The charge state of the double dot is controlled with gate voltages V_L and V_R. These gate voltages are swept to change the energy difference of the qubit state, ε (same definition as in Sect. 2.1). To test the feasibility of detecting charge states of the DQD, dc conductance measurement is carried out for the QPC electrometer. The QPC conductance is adjusted at the maximal sensitivity condition (about half of the quantized conductance). In order to improve the signal to noise ratio, the charge state of the DQD is modulated by applying low-frequency (100 Hz) sinusoidal voltage, V_{mod}, to the electrode as shown in Fig. 5(a), and the corresponding modulated current I_{det} through the QPC is measured with a lock-in amplifier.

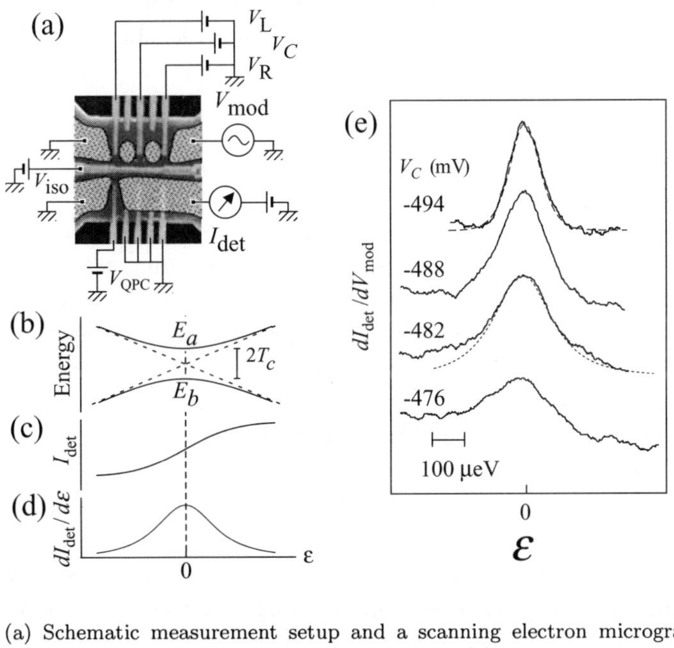

Fig. 5. (a) Schematic measurement setup and a scanning electron micrograph of a double quantum dot coupled with a QPC electrometer. (b) Schematic energy diagram of eigenstates. (c) Current through the QPC. (d) The derivative of the QPC current. (e) Experimental charge detection signal. The dashed and dotted lines are fitted to the data by assuming thermal distribution and coherent tunneling coupling, respectively.

Figure 5(b) shows a schematic energy diagram of eigenenergies E_a and E_b when the energy bias ε is changed. The QPC current would be influenced by the charge occupation, $I_{det} \sim |\langle R|\varphi\rangle|^2 - |\langle L|\varphi\rangle|^2$, as schematically shown for the bonding state in Fig. 5(c). The derivative curve, $dI_{det}/d\varepsilon$, is shown in Fig. 5(d), and is given by

$$\frac{dI_{det}}{d\varepsilon} \sim \frac{\hbar^2 T_C^2}{(\hbar^2 T_c^2 + \varepsilon^2)^{3/2}}. \tag{4}$$

Figure 5(e) shows experimental traces of dI_{det}/dV_{mod} around the charge degeneracy point $\varepsilon \equiv E_L - E_R = 0$, in which V_L and V_R are swept simultaneously in the opposite direction. The energy scale in the figure was determined from photon assisted tunneling spectroscopy with microwave irradiation (not shown). The peak in the topmost trace at the central gate voltage V_C = -494 mV shows the narrowest peak observed in this measurement. This broadening is probably related to the external noise or ac modulation voltage used in this experiment. The profile can be fitted well by assuming Fermi-Dirac distribution of charges as shown by the dashed line,

$$\frac{dI_{\text{det}}}{d\varepsilon} \sim \cosh^{-2}(\varepsilon/k_B T_{eff}), \tag{5}$$

with the effecitive thermal energy of $k_B T_{eff} \sim 30$ μeV as a parameter. As the central gate voltage V_C is made less negative to increase the tunneling coupling, the peak broadens, suggesting that the two dots are coherently coupled by the tunneling coupling. The peaks for the lowest two traces in Fig. 5(e) can be fitted well with Eq. 4 (see dotted line fitted with $\hbar T_c \sim$ 60 μeV for V_C = -482 mV). The disagreement on the left hand side arose from the broad positive signal from another peak on the left.

The experiment demonstrates that the QPC detector can determine the charge distribution in a double dot. More accurate parameters may be obtained by taking into account the thermal distribution of ground and excited states [33]. A high-frequency technique would allow us to determine instantaneous charge state and provide single-shot charge readout scheme [32].

3. Electron spin qubit

3.1. *Handling single spins in nanostructures*

Electron-spin based quantum computation is motivated by its expected long decoherence time.[13,34] For instance, electron spin bound to a donor in silicon shows phase relaxation time of $T_2 \sim 300$ μs.[35] Coherent manipulation of electron spins has been studied in many systems with ensemble of spins. However, little work has been done for single electron spins. In order to control each electron spin (qubit) independently, techniques for manipulating and measuring single spins are essential.

A simple idea for one-qubit operation is to exploit the electron spin resonance (ESR).[37] Electron spin in a static magnetic field precesses about the field axis, which can be considered to act as a phase-shift gate for the spin qubit. A radio-frequency magnetic field applied perpendicular to the static magnetic field induces Rabi oscillation at the ESR condition, which can be considered to act as a rotation gate. Therefore, arbitrary qubit manipulation can be performed by controlling the rf magnetic field. In order to control each single qubit independently, the ESR frequency can be made different for different qubits so as to prepare frequency-selective spin qubits. For instance, the effective g-factor of each electron spin can be made different for different QDs by using g-factor engineering,[13] or a moderate magnetic field gradient can be used to change the Zeeman splitting energy. However, since magnetic dipole coupling is very weak, a typical one-qubit operation with ESR would require a relatively long time ($\gtrsim$100 ns) for a rotation gate with an experimentally accessible rf magnetic field.

There are some other schemes available to control spins in a shorter time. Owing to the spin-selection rule in the optical transition between the valence and conduction band of a semiconductor, spin can be effectively manipulated using the optical Stark effect under off-resonant optical excitation.[38] The g-factor is actually anisotropic in an anisotropic confinement potential, and it can be expressed by a tensor. The g-tensor can be modulated in a tailored heterostructure with an electric field, which allows spin resonance with a microwave electric field.[39] Spin-orbit coupling for electrons in an asymmetric confinement potential induces spin precession when the electron travels along one direction.[47,40] These mechanisms have been mainly discussed for 2D electron systems. They can be applied to electron spins in QDs in principle. As discussed in the next subsection, spin-orbit coupling can be used to convert an orbital state (charge qubit) to a spin state, and thus a spin state can be manipulated with a charge qubit.

Two-qubit operation can be performed by using the exchange interaction between two spins, one in each dot.[36] The exchange energy can be controlled by changing the overlap of the wavefucntions confined in the dots. As the name suggests, the exchange interaction can be used to exchange the quantum information between the two electron spins; that is, it provides a SWAP gate for swapping the qubit information. The exchange energy in a single QD is now well understood from excitation spectra in a magnetic field.[41] However, the exchange energy in a DQD has not been studied well because the energy is usually too small to be resolved in conventional excitation spectra.[42] Slightly different schemes combined with spin-dependent tunneling etc. would be required to search for exchange splitting in a double dot. An electron pair in a spin triplet (singlet) state in a DQD cannot (can) occupy the same orbital in one dot, a situation known as the Pauli spin blockade.[43] Recent spin blockade measurement in the cotunneling regime has detected a small exchange splitting comparable to the thermal energy.

Difficulties in experimental studies for single spin state arises from the extremely small sensitivity. The conventional microwave transmission/reflection measurement for detecting ESR signal cannot be applied to detect a single spin. However, the recent progress in single spin detection is impressive. For instance, ESR signal from a single spin of a defect in Si can be barely detected by a current passing through a nearby defect.[44] Actually, single-shot readout for a spin qubit has been proposed and demonstrated by spin-selective transport detected with a QPC charge detector.[45,46]

3.2. *Spin-orbit interactions*

Electron spins in the conduction band of a semiconductor are influenced by the periodic potential of the semiconductor lattice. According to the standard $k \cdot p$ perturbation theory, electrons having a momentum away from the symmetric point have a small fraction of p-like orbitals that exhibit spin-orbit coupling.[47] This gives rise to two important effects on the conduction band electrons. Spin-up and spin-down states are energetically separated in the presence of an external magnetic field to the opposite direction of the bare Zeeman splitting. This effect is usually absorbed in the effective g-factor ($g^* = -0.44$ for GaAs), which becomes less than 2. The other effect appears as spin-orbit coupling for electrons in the conduction band, provided that the potential for the electron has a space inversion asymmetry. Compound semiconductors, such as GaAs, have an intrinsic bulk inversion asymmetry (BIA) in the crystal lattice, and the confinement potential of a QD may have structural inversion asymmetry (SIA) in some devices. These asymmetries induce spin-orbit coupling, which has been investigated in various experiments, like the measurement of weak anti-localization effect in diffusive 2D conductor.[48]

For an electron confined in the 2D harmonic potential of QD disk characterized by radial quantum number n, angular quantum number l, and spin quantum number s, the BIA spin-orbit interaction couples two states having different quantum numbers $\Delta l = \Delta s = \pm 1$ to the lowest order, while the SIA interaction couples two states with $\Delta l = -\Delta s = \pm 1$.[49] If the dot potential is not well approximated by the 2D harmonic potential, the selection rule should be smeared out and spin-orbit coupling should appear between any states with $\Delta s = \pm 1$. For simplicity, we consider two orbitals, a and b, and the spin-orbit coupling energy

$$\Delta_{so} = |\langle \psi_{a,\uparrow} | H_{so} | \psi_{b,\downarrow} \rangle| = |\langle \psi_{a,\downarrow} | H_{so} | \psi_{b,\uparrow} \rangle|. \tag{6}$$

This interaction mixes the spin-up and spin-down electron wavefunctions, which can be written as

$$\begin{aligned} \tilde{\psi}_{a,\uparrow/\downarrow} &\approx \psi_{a,\uparrow/\downarrow} - \frac{\Delta_{so}}{\varepsilon_{ab} \pm E_Z/2} \psi_{b,\downarrow/\uparrow} \\ \tilde{\psi}_{b,\uparrow/\downarrow} &\approx \psi_{b,\uparrow/\downarrow} + \frac{\Delta_{so}}{\varepsilon_{ab} \mp E_Z/2} \psi_{a,\downarrow/\uparrow}. \end{aligned} \tag{7}$$

Here, $\psi_{a,\uparrow/\downarrow}$ ($\psi_{b,\uparrow/\downarrow}$) is the unperturbed wavefunction of a spin-up/down electron in the orbital a (b), and $\varepsilon_{ab} = E_b - E_a$ is the energy difference of orbitals a and b. When magnetic field B is applied, each spin sublevels are separated by Zeeman energy $E_Z = |g^*|\mu_B B$. If the magnetic field is such that the spin-down branch of orbital a is energetically equal to the spin-up branch of orbital b, each sublevel is repelled from one another due to spin-orbit coupling as schematically shown in Fig. 6. Since ε_{ab} can be slightly changed by deforming the confinement potential with gate voltages, the crossing magnetic field B_X can be adjusted by the gate voltages. Therefore, coherent oscillation is expected between the two states. Note that the spin and orbital degrees of freedom are simultaneously manipulated in this case.

Although spin-orbit interactions just couple spin and orbital degrees of freedom, they introduce decoherence paths to the spin qubit via the orbital degree of freedom from various kinds of environments, some of which have been discussed in Sect. 2.2. According to theoretical predictions, phonon emission is a dominant decoherence mechanism in the presence of spin-orbit coupling.[51] Mixing different spins induces phonon transition between these states. After introducing measurement technique in the next subsection, we discuss our experimental studies on this subject in Sect. 3.4.

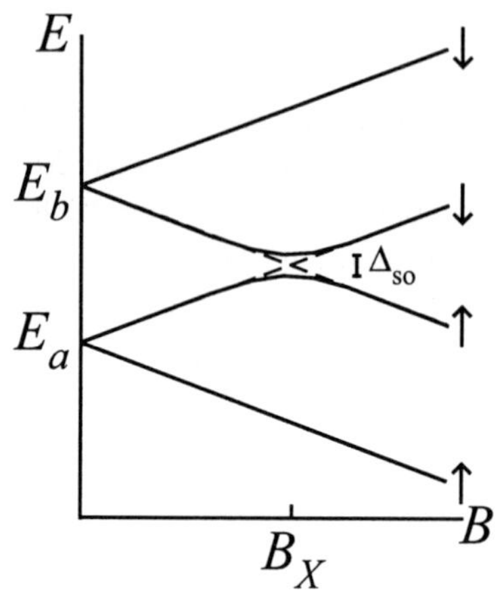

Fig. 6. Schematic energy diagram of anti-crossing behavior under spin-orbit coupling.

3.3. *Electrical pump and probe experiment*

Energy relaxation measurement is a fundamental way to investigate how strongly a quantum system is coupled to the environment. Here, we introduce electrical pump and probe methods for investigating energy relaxation time in a QD. A typical QD device has source (S), drain (D) and gate (G) electrodes to control the transport through the dot as shown in Fig. 7(a). The relaxation time can be measured by applying a time-dependent gate voltage, $V_g(t)$, with a single or double step waveform. We consider energy relaxation process from a first excited state (ES) to the ground state (GS) of an N electron QD. Each state is coupled to the left and right lead with tunneling rates Γ_L and Γ_R, respectively. We consider strongly asymmetric barriers, $\Gamma_L \gg \Gamma_R$, where an electron is injected from the thinner barrier (Γ_L) and ejected through the thick barrier (Γ_R) under a small positive bias voltage in the saturated transport regime.[53]

In the single-step pulse scheme, the gate voltage $V_g(t)$ is switched between low voltage V_l and high measurement voltage V_m as shown in Fig. 7(b). First, we prepare the QD in the $N-1$ Coulomb blockade region by setting the gate voltage to $V_g = V_l$ [Fig. 7(c)]. In this situation, both the N-electron ES and the GS are energetically located above the chemical potential of the two electrodes, μ_L and μ_R, and thus become empty. Then, the gate voltage is switched to V_m, where only the N-electron ES is located between μ_L and μ_R (the transport window), as shown in Fig. 7(e). First, an electron enters either the N-electron ES or the GS, but only one electron

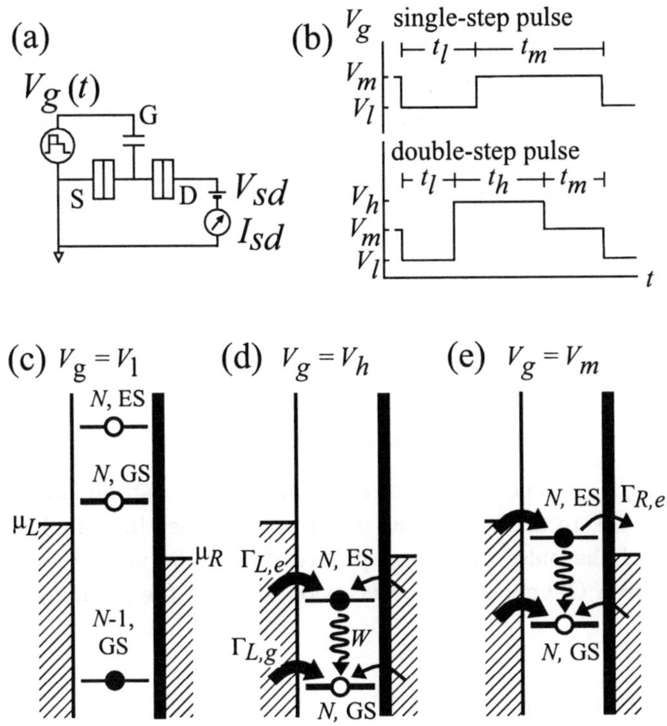

Fig. 7. (a) Schematic diagram for electrical pump and probe measurement. (b) The pulse waveform for single- and double-step voltage pulses. (c-e) Energy diagram for initialization (c), relaxation (d), and measurement (e).

can enter the dot at a time. If an electron enters the ES, it may relax to the GS or tunnel to the right electrode to contribute to the net current. However, the transport is blocked once the GS is occupied (Coulomb blockade). Because of $\Gamma_L \gg \Gamma_R$, an electron can stay in the ES for a relatively long time $\sim \Gamma_R^{-1}$, during which relaxation may take place. The transport through the ES is terminated by the relaxation process. Therefore, transport is sensitive to the relaxation time. This simple method can be applied to measure the relaxation rate W in the range of $\Gamma_R \lesssim W \lesssim \Gamma_L$.

An improved method to determine a longer relaxation time, i.e., $W < \Gamma_R$, is to use a double-step voltage pulse, in which V_g is switched between three voltages, V_l, V_h and V_m, as shown in Fig. 7(b).[53] Initially, when $V_g = V_l$ [Fig. 7(c)], empty N states are prepared during the period t_l. When V_g is increased to V_h [Fig. 7(d)] such that both N-electron ES and

GS are located below the chemical potentials, an electron enters either the ES or the GS. Once an electron enters either state, it can not leave the dot, nor can another electron enter it (Coulomb blockade). The dot is now effectively isolated from the electrodes. If an electron has entered the ES, the ES may relax to the GS during this period. Therefore, the average electron number in the ES after the period t_h is approximately given by

$$\rho_e = A \exp(-W t_h) \tag{8}$$

for $t_h \gg \Gamma_L^{-1}$, where $A \simeq \frac{\Gamma_{L,e}}{\Gamma_{L,e}+\Gamma_{L,g}}$. When V_g is switched to V_m [Fig. 7(e)], the electron, if it has not yet relaxed to the GS, can tunnel out to the drain. The average number of electrons for net current during a period n_t can be written as

$$n_t = A' \exp(-W t_h), \tag{9}$$

where $A' = \Gamma_{R,e}/\Gamma_{R,g}$. Therefore, the relaxation rate W can be obtained by measuring $n_t(t_h)$. In this scheme, an electron is pumped into the ES on the rising edge of the pulse to V_h and probed during the middle voltage V_m. The N-electron QD experiences relaxation between the pumping and probing sequence.

3.4. *Spin relaxation time in a QD*

The energy relaxation time from a first excited state to the ground state observed in a few-electron vertical QD is summarized in Fig. 8, in which two kinds of relaxation processes are compared. The momentum relaxation process appearing in an $N = 1$ QD takes place with electron phonon interaction. For energy spacing in the meV region, spontaneous emission of a *phonon* dominates the relaxation process at low temperature [17]. The energy spacing coincides with the acoustic-phonon energy in the linear dispersion regime. Actually, the phonon emission rate calculated from Fermi's golden rule, considering both deformation and piezoelectric types of coupling, reproduces the experimental data as shown by the solid line in Fig. 8.

In contrast to the relaxation time of about 10 ns for an $N = 1$ QD, the relaxation time is remarkably different for an $N = 2$ QD (solid circles in Fig. 8). At low magnetic fields (less than 2.5 T for this sample), the many-body ground state is a spin-singlet, while the first excited state is a spin-triplet.[8,41] Because of direct and exchange Coulomb interactions, the energy spacing between the two states ($\sim$0.6 meV at $B = 0$ T), is smaller than the energy spacing ($\sim$2.5 meV) for $N = 1$ QD. Energy relaxation from the triplet state to the singlet state shows relaxation time of about 200 μs,

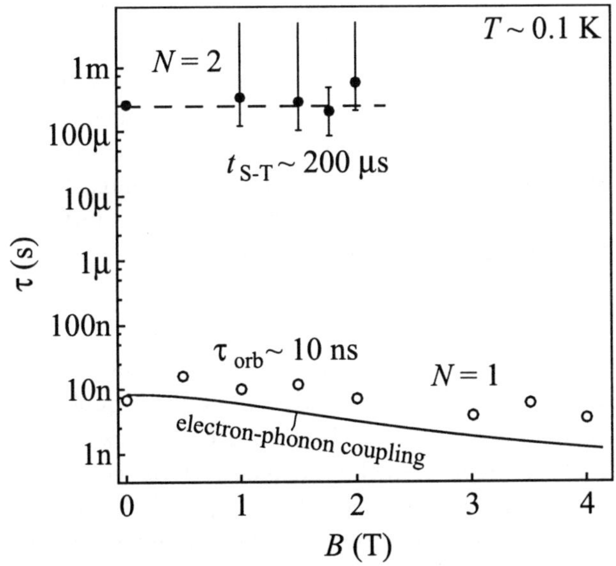

Fig. 8. Energy relaxation time from a first excited state to the ground state in a quantum dot.

that is more than four orders of magnitude longer than that for orbital relaxation process. A simple phonon-emission transition from the spin-triplet to the spin-singlet is forbidden by spin conservation. Actually, the relaxation time in this case is determined by inelastic cotunneling, in which an electron spin in the dot is exchanged with another electron spin with opposite direction in the lead. Since cotunneling effect can be eliminated by isolating the dot from the electrodes, the intrinsic spin relaxation time for the present case should be longer than 200 μs. The large difference in the relaxation time means that the spin degree of freedom is well isolated from the orbital degree of freedom in a QD. Spin relaxation mechanisms, such as spin-orbit and hyperfine interactions,[51] must have had only a weak effect on the breaking of the "forbidden" symmetries. Spin-orbit interactions are predicted to make the dominant contribution to spin relaxation in GaAs QD systems, although this is still an extremely small effect. According to the perturbative approach for the two-electron case similar to Eq. 6, we can deduce an upper bound of spin-orbit coupling energy $\Delta_{so} < 4$ μeV.[52]

Our experiments indicate that the spin degree of freedom in QDs is well isolated from the orbital degree of freedom. This is attractive for applications to spin memories and spin qubits. The energy relaxation time (T_1)

of a single-electron spin qubit in a static magnetic field can be estimated using Eq. 7. Since $\Delta_{so} < 4\ \mu$eV, this yields $T_{1,so} > 1$ ms for a Zeeman splitting $\varepsilon_Z \sim 0.1$ meV and $\varepsilon_{1s-2p} \sim 1.2$ meV at $B \sim 5$ T. Recent measurement of the energy relaxation time between Zeeman sublevels in an $N = 1$ QD shows $T_1 \gtrsim 1$ ms,[46,54] which supports our estimation. These relaxation times are much longer than the time required for typical one- and two-qubit operations and therefore encouraging for further research in the use of the spin degree of freedom in QDs.

4. SUMMARY

We have discussed electron charge qubits and spin qubits in semiconductor QDs. The current status and prospects of these qubits for quantum information devices can be summarized using DiVincenzo criteria checklist as shown in Fig. 9. Semiconductor nanofabrication technologies are well matured in the fabrication of large-scale integrated classical circuits. The requirement for fabricating quantum circuits is to remove unwanted fluctuation or decoherence in the qubit performance.

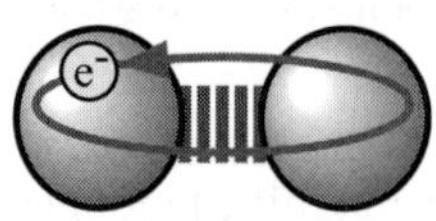

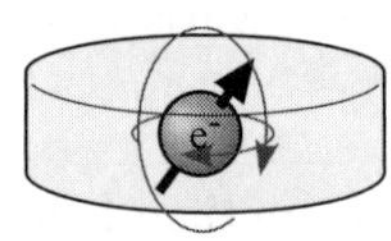

◎ confirmed experimentally ○ expected △ problematic	Electron-charge qubit	Electron-spin qubit
scalable quantum bit	○ Semiconductor nanofabrication technologies (EB litho, self-organized, molecules, SPM related, etc.)	
Initialization	◎ Electron injection [12]	◎ Electron injection [46]
One-qubit operation	◎ Tunneling coupling (volt. pulse) [12]	○ Electron spin resonance [37] Spin orbit coupling etc.
Two-qubit operation	○ Dipole coupling (volt. pulse) [11]	○ Exchange coupling [36]
Readout capability	◎ Tunneling current [12]	◎ Tunneling current [46]
Single shot meas.	○ SET/QPC electromenter [32]	◎ QPC electromenter [46]
Phase relax. time	△ Charge fluctuation [12, 22]	○ expected to be long, but not clear
Energy relax. time	△ Phonon emission [12, 26]	◎? phonon emission (spin-orbit coupling) [25, 46, 54]

Fig. 9. DiVincenzo check list for electron charge qubit and electron spin qubit.

As for electron charge qubits, we have demonstrated the initialization, measurement, and full one-qubit manipulation by applying a voltage pulse to an electrode as shown in Sect. 2.1. Electron injection is an effective way to prepare a well-defined state of the qubit. Controlling the energy bias of the qubit with a tailored pulse waveform is used to perform phase shift gate and rotation gate consecutively. Standard current measurement is useful for probing ensemble characteristics of the coherent oscillations. And single-shot measurement will be available with a sensitive charge detector as described in Sect. 2.3. A two-qubit gate is expected using electrostatic dipole coupling between two charge qubits.[11] However, relatively short relaxation times are introduced from background charge fluctuations and electron-phonon coupling.

Longer relaxation time is anticipated in electron spin qubits. We have found that electron spin in a QD actually has a long energy relaxation time even in a polar material. However, the phase relaxation time of an electron spin in a QD is not yet clearly understood experimentally. Although coherent spin manipulations have been discussed with ensemble spins, experimental techniques for detecting single electron spin with a reasonable time resolution are needed. While standard ESR is useful for pure spin manipulation, alternative schemes associated with spin orbit effects in semiconductors are promising for faster manipulations. Exchange coupling in a double dot, which is important for two-qubit operation, will be explored using various techniques.

Acknowledgments

This work was partly supported by a Grant-in-Aid for Scientific Research from the Japan Society for the Promotion of Science.

References

1. M. A. Nielsen and I. L. Chuang, Quantum Computation and Quantum Information, (Cambridge, 2000).
2. Eds. D. Bouwmeester, A. Ekert, and A. Zeilinger, The Physics of Quantum Information, (Springer, 2000).
3. L. M. K. Vandersypen, M. Steffen, G. Breyta, C. S. Yannoni, M. H. Sherwood, and I. L. Chuang, Nature 414, 883 (2001).
4. A. Rauschenbeutel, G. Nogues, S. Osnaghi, P. Bertet, M. Brune, J.-M. Raimond, S. Haroche, Science 288, 2024 (2000).
5. C. J. Hood, T. W. Lynn, A. C. Doherty, A. S. Parkins, H. J. Kimble, Science 287, 1447 (2000).
6. A. Wallraff, D. I. Schuster, A. Blais, L. Frunzio, R.- S. Huang, J. Majer, S. Kumar, S. M. Girvin, R. J. Schoelkopf, Nature 431, 162 (2004).
7. L. P. Kouwenhoven, D. G. Austing, and S. Tarucha, Reports on Progress in Physics **64**, 701 (2001).

8. S. Tarucha, D. G. Austing, T. Honda, R. J. van der Hage and L. P. Kouwenhoven, Phys. Rev. Lett. **77**, 3613 (1996).
9. T. H. Oosterkamp, T. Fujisawa, W. G. van der Wiel, K. Ishibashi, R. V. Hijman, S. Tarucha, and L. P. Kouwenhoven,Nature 395, 873 (1998).
10. D. P. DiVincenzo, in *Mesocopic Electron Transport* edited L. L. Sohn, L. P. Kouwenhoven, and G. Schön, NATO ASI series E 345, Kluwer Academic, Dordrecht (1997), pp. 657-677.
11. A. Barenco, D. Deutsch, A. Ekert, and R. Jozsa, Phys. Rev. Lett. 74, 4083 (1995).
12. T. Hayashi, T. Fujisawa, H. D. Cheong, and Y. Hirayama, Phys. Rev. Lett. 91, 226804 (2003).
13. D. Loss and D. P. DiVincenzo, Phys. Rev. A 57, 120 (1998).
14. J. M. Taylor, C. M. Marcus, and M. D. Lukin, Phys. Rev. Lett. 90, 206803 (2003).
15. F. R. Waugh, M. J. Berry, D. J. Mar, R. M. Westervelt, K. L. Campman, and A. C. Gossard, Phys. Rev. Lett. 75, 705 (1995).
16. C. Livermore, C. H. Crouch, R. M. Westervelt, K. L. Campman, and A. C. Gossard, Science 274, 1332 (1996).
17. T. Fujisawa, T. H. Oosterkamp, W. G. van der Wiel, B. W . Broer, R. Aguado, S. Tarucha, and L. P. Kouwenhoven, Science 282, 932 (1998).
18. J. M. Elzerman, R. Hanson, J. S. Greidanus, L. H. Willems van Beveren, S. De Franceschi, L. M. K. Vandersypen, S. Tarucha, and L. P. Kouwenhoven, Phys. Rev. B 67, 161308 (2003).
19. T. Fujisawa, T. Hayashi, Y. Hirayama, J. Vac. Sci. Technol. B 22, 2035 (2004).
20. T. Fujisawa, T. Hayashi, H. D. Cheong, Y. H. Jeong, Y. Hirayama, Physica E 21, 1046 (2004).
21. E. Paladino, L. Faoro, G. Falci, and Rosario Fazio, Phys. Rev. Lett. 88, 228304 (2002).
22. S. W. Jung, T. Fujisawa, Y. Hirayama and Y. H. Jeong, Appl. Phys. Lett. 85, 768 (2004).
23. M. J. Kirton and M. J. Uren, Advances in Physics 38, 367 (1989).
24. M. Eto, Jpn. J. Appl. Phys. 40, 1929 (2001).
25. T. Fujisawa, D. G. Austing, Y. Tokura, Y. Hirayama, and S. Tarucha, Nature 419, 278 (2002).
26. T. Brandes and B. Kramer, Phys. Rev. Lett. 83, 3021 (1999).
27. Seeger, K., *Semiconductor Physics: An Introduction* 153-213 (Springer-Verlag, Berlin, 1985).
28. S. Gardelis, C. G. Smith, J. Cooper, D. A. Ritchie, E. H. Linfield, Y. Jin, and M. Pepper, Phys. Rev. B 67, 073302 (2003).
29. W. Lu, Z. Ji, L. Pfeiffer, K. W. West, A. J. Rimberg, Nature 423, 422 (2003).
30. T. Fujisawa, T. Hayashi, Y. Hirayama, H. D. Cheong and Y. H. Jeong, Appl. Phys. Lett. 84, 2343 (2004).
31. R. J. Schoelkopf, P. Wahlgren, A. A. Kozhevnikov, P. Delsing, D. E. Prober, Science, 280, 1238 (1998).
32. A. Aassime, G. Johansson, G. Wendin, R. J. Schoelkopf, and P. Delsing, Phys. Rev. Lett. 86, 3376 (2001).

33. L. DiCarlo, H. J. Lynch, A. C. Johnson, L. I. Childress, K. Crockett, C. M.
34. V. Cerletti, W. A. Coish, O. Gywat and D. Loss, cond-mat/0412028. Marcus, M. P. Hanson, and A. C. Gossard, Phys. Rev. Lett. 92, 226801 (2004).
35. M. Chiba and A. Hirai, J. Phys. Soc. Jpn. 33,730 (1972).
36. G. Burkard, D. Loss and D. P. DiVincenzo, Phys. Rev. B 59, 2070 (1999).
37. H.-A. Engel and Daniel Loss, Phys. Rev. Lett. 86, 4648 (2001).
38. J. A. Gupta, R. Knoble, N. Samarth, and D. D. Awschalom, Science 292, 2458 (2001).
39. Y. Kato, R. C. Myers, D. C. Driscoll, A. C. Gossard, J. Levy, D. D. Awschalom, Science 299, 1201 (2003).
40. Y. Kato, R. C. Myers, A. C. Gossard, and D. D. Awschalom, Nature 427, 50 (2004).
41. S. Tarucha, D. G. Austing, Y. Tokura, W. G. van der Wiel, and L. P. Kouwenhoven, Phys. Rev. Lett. **84**, 2485 (2000).
42. Vitaly N. Golovach, Daniel Loss, Phys. Rev. B 69, 245327 (2004).
43. K. Ono, D. G. Austing, Y. Tokura, and S. Tarucha, Science 297, 1313 (2002).
44. M. Xiao, I. Martin, E. Yablonovitch, H. W. Jiang, Nature 430, 435 (2004).
45. B. E. Kane, N. S. McAlpine, A. S. Dzurak, R. G. Clark, G. J. Milburn, He Bi Sun, and H. Wiseman, Phys. Rev. B 61, 2961 (2000).
46. J. M. Elzerman, R. Hanson, L. H. Willems van Beveren, B. Witkamp, L. M. K. Vandersypen, L. P. Kouwenhoven, Nature 40, 431 (2004).
47. R. Winkler, Spin-orbit coupling effects in two-dimensional electron and hole systems, Springer.
48. J. B. Miller, D. M. Zumbuhl, C. M. Marcus, Y. B. Lyanda-Geller, D. Goldhaber-Gordon, K. Campman, and A. C. Gossard, Phys. Rev. Lett. 90, 076807 (2003).
49. C. F. Destefani, S. E. Ulloa, and G. E. Marques, cond-mat/030702.
50. D. V. Bulaev and D. Loss, cond-mat/0409614.
51. A. V. Kaetskii, and Yu. V. Nazarov, Phys. Rev. B 61, 12639 (2000).
52. T. Fujisawa, D. G. Austing, Y. Tokura, Y. Hirayama, S. Tarucha, J. Physics: Cond. Matt. 33, R1395 (2003).
53. T. Fujisawa, D. G. Austing, Y. Hirayama, and S. Tarucha, Jpn. J. Appl. Phys. 42, 4804 (2003).
54. M. Kroutvar, Y. Ducommun, D. Heiss, M.Bichler, D. Schuh, G. Abstreiter, and J. J. Finley, Nature 432, 81(2004).

Superconducting Quantum Computing: Status and Prospects *

F.K. Wilhelm

Department Physik, Center for Nanoscience, and Arnold-Sommerfeld Center for Theoretical Physics
Ludwig-Maximilians-Universität
Theresienstr. 37
D-80333 München, Germany
E-mail: wilhelm@theorie.physik.uni-muenchen.de

K. Semba

NTT Basic Research Laboratories, NTT Corporation,
Atsugi, Kanagawa 243-0198 Japan
CREST, Japan Science and Technology Agency
4-1-8 Honcho, Kawaguchi, Saitama 332-0012, Japan
E-mail: semba@nttbrl.jp

We review the experimental and theoretical status of superconducting quantum bits based on Josephson junctions in view of DiVincenzo's criteria.

This article takes a momentary (summer 2005) snapshot of the field of superconducting Josephson qubit. It serves on the one hand as an introduction to the general concept aimed at a broader quantum computing audience containing a guide to the more specifit literature. A particular focus is given to the rather unique idea of manipulating macroscopic states quantum-coherently and the intellectual and practical challenges and implications of that work. On the other hand, we have attempted to review the most significant results of the field. Such a choice is necessarily subjective and we are probably not fully exhaustive. For exemplifying details we have typically described our own work — this is what we know best — in a way which highlights the broader ideas and should enable the reader to comprehend related work of other authors from the original literature.

* Dedicated to the memory of Martti Salomaa. Parts of the text are based on the Habilitationsschrift of FKW, Ludwig-Maximilians-Universität, 2004.

1. Why Josephson qubits?

The way of computing as we know it in today's information-processing devices is called *classical computing.* This implies, that the binary information stored and manipulated is purely classical in nature: A bit is exclusively in one of its fundamental states 0 and 1. Computer programs are deterministic, such that (in principle) operating a program on a fixed set of data leads to the same output data every time. The enormous progress in hardware improvement follows the self-fulfilling Moore's law [1], claiming that computer hardware performance will double every 18 months. Although failures of Moore's law are periodically predicted and have to occur at some time due to the laws of nature, it is by now expected to still persist for a time of 10–20 years [2,3].

Quantum information, the type of information processed in a quantum computer, is radically different [4,5]: It uses the predictions of quantum mechanics, most notably the possibility of interfering possibilities for different values of observables at the same time, and to distribute the information on the physical state nonlocally using entangled states. These properties can be used as a computational resource. Using superpositions allows to operate an algorithm on all possible input values at the same time (massive quantum parallelism), using entangled states and measurements allows to act on all qubits (quantum bits) simultaneousely.

The requirements for building a universal quantum computer have been collected very early by DiVincenzo [6,7]. These five criteria enjoy broad recognition and are, supplemented by two more on quantum communication, the basis for contemporary quantum computing research programs [8]. One has to be aware, that this standard paradigm is by no means exclusive, in fact, they are based on the circuit model of quantum computing, which does not apply to, e.g., adiabatic algorithms [9].

The DiVincenzo criteria in a timely formulation require:

(1) A scalable physical system of well-characterized qubits
(2) The ability to initialize the state of the qubits to a simple fiducial state
(3) Long (relative) decoherence times, much longer than the gate-operation time
(4) A universal set of quantum gates
(5) A qubit-specific measurement capability
(6) The ability to interconvert stationary and flying qubits
(7) The ability to faithfully transmit flying qubits between specified locations

This report will describe how superconducting qubits perform on these criteria.

So far, the physical implementation of quantum computing is most successful in nuclear magnetic resonance (NMR), where seven qubits have been implemented [10]. Remarkably, in NMR there is no "strong" measurement capability nor the option to initialize a well-defined initial state [11]. Another successful line of implementations comes from optical and atomic physics, such as ion traps [12,13], atoms in cavities [14,15], linear optics [16], and neutral atoms [17]. All these realizations have the general idea in common, that they start out from microscopic quantum systems, systems whose quantum-mechanical properties are well-established and in principle easy to demonstrate experimentally. Consequently, the phase coherence times are very long. The experimental progress towards today's level has been the ability to externally control these systems with high accuracy and to connect many of them to larger coupled quantum systems. However, it is still not evident whether these systems are really scalable, although a number of theoretical proposals for scalable ion trap computing have been brought forward [18].

2. Solid-state quantum computation

Solid state circuits are readily scalable to nearly arbitrary sizes, as can be seen in the computers available today. In these computers, the information being processed is purely classical. The shear size of a solid renders many of the quantum effects invisible, which would be visible in atomic and molecular systems. However, the physics laying the foundation of classical computers is already fundamentally quantum: The function of transistors relies on the band structure of semicondcutor materials [19], which is a generic quantum-mechanical effect. Remarkably, also the size of a single transistor (characterized by the gate length) is on the order of 100 nm or even below and hence does not rule out mesoscopic effects [20]. The same holds for the enormous clock speed of several GHz. Hence, as solid state setups are more and more miniaturized, quantum effects become important. This is the idea of mesoscopic solid-state physics [21,22]. Superconducting mesoscopic systems are in particular appealing, as the conduction electrons are condensed into a macroscopic quantum wave function already. It has already been proposed in the 80s (for SQUID systems) [23–27]and in the 90s (for Coulomb blockade devices) [28,29], that collective variables such as flux and charge can be brought into superposition of two macroscopically distinct values. This enterprise has been pursued ever since, but it was only in the late 90s that it received the high degree of attention and interest it has now. This increase in interest was largely due to the perspective to realize a scalable solid-state quantum computer [30].

The reason why the *collective* variables of solid state systems usually behave classically is decoherence and the lack of quantum fluctuations which allow the preparation of generically non-classical states. As will be detailed in a later section, the reason for this is coupling to an environment with many degrees of freedom and low-lying excitations. This is *the* main challenge in designing solid-state qubits. A number of proposals has been brought forward, which can be roughly classified in two classes: Spin quantum computing using a controlled exchange interaction such as in Phosphorus in Silicon [31] and spins in quantum dots [32], and pseudospoin quantum computing, where two-state systems other than spin are used such as charge states in quantum dots and superconducting quantum bits. The former promise very long coherence times but are difficult to fabricate and readout, thus, the experimental realizations are on a rather pioneering stage [33,34], but the possibilities are enormous [35–37]. The status of the latter will be detailed in the following sections of this chapter.

3. Superconductivity and the Josephson effect

One specific class of promising qubit implementations is based on superconducting Josephson junctions.

Superconductivity has been discovered already in 1911 [38]. It manifests itself by vanishing electrical DC-resistance and by perfect diamagnetism.

The theoretical understanding of superconductivity is now very mature [39–42]. Already the phenomenological London theory [43] of superconducting electrodynamics describes many superconducting phenomena rather well. Its modern formulation can be derived microscopically and accounts for the fact that superconducting phenomena can be described by the dynamics of a single particle-like wavefunction which describes the collective properties of *all* superconducting electrons. In the presence of an attractive effective interaction, the conduction electrons form Cooper pairs, which condense in momentum space into a collective ground state. Thus, the "superconducting electrons" of the phenomenological theories are Cooper pairs whose charge is twice the elementary charge. The attractive interactions between electrons in metallic superconductors is phonon-mediated. In superconducting metals, this indirect interaction dominates over the Coulomb repulsion, which is screened and whose phase space is usually restricted by the Pauli principle.

Superconductivity is in itself a macroscopic quantum phenomenon: The simple manifestations of superconductivity such as flux quantization [44] and persistent currents are quantum properties of the condensate wavefunction. This wavefunction is occupied by a macroscopic number of particles. However, in bulk superconductors the wavefunction itself is well-defined and the collective variables, number and phase, do not have quantum uncertainties. The elementary excitation of superconductors are quasiparticles. They are separated from the condensate by an an energy gap Δ. This can be identified with the order-parameter of Ginzburg-Landau theory [†].

The gap of the elementary excitations makes superconductors attractive for solid-state quantum computation: The elementary electronic excitations are costly in energy and can be supressed at low temperatures. At these low temperatures, also the lattice vibrations are frozen out. Thus, conventional superconductors promise to have very low intrinsic decoherence [45] although phonons may play a role [46].

The Josephson effect [47] is recognized as a hallmark of superconductivity. Its basic statement is that Cooper pairs can coherently tunnel between two superconductors connected by a weak link [48,49]. This gives rise to a supercurrent which is controlled by the difference of the phases of the order parameter in the two superconductors $\phi = \phi_1 - \phi_2$. For weak coupling between the superconductors, which will be assumed henceforth unless stated otherwise [50], the current-phase-relation is sinusoidal,

$$I = I_{\rm c} \sin \phi. \tag{1}$$

This is the first Josephson equation. From basic consideration of gauge invariance, one can derive the second Josephson relation, which connects the time evolution of the phase difference ϕ with the difference in chemical potential of the Cooper pairs

$$\hbar \dot{\phi} = 2eV. \tag{2}$$

The Josephson effect has first been microscopically dervied from the BCS theory and the tunneling Hamiltonian by Josephson. Later, a number of pedagogical derivations have appeared, which do *not* contain details of BCS theory and hence outline that the Josephson effect is a universal phenomenon whenever two spatially separated coherent wave functions are connected by a weak link [27,51,52]. In fact, Josephson effects have been observed

---

[†] In some cases, such as superconductors with magnetic impurities or high-temperature-superconductors close to surfaces, one still keeps an order parameter Δ, although excitations below Δ exist.

in systems like superfluid Helium 3 and 4 [53,54], Bose-Einstein-condensates [55], and molecular junctions [56].

Real Josephson junctions contain more possibilities to transport charge than the supercurrent. The main mechanisms are:

i) A resistive channel formed by transport of quasiparticles. In tunnel junctions, this contribution is strongly gapped. Thus, at low T and V, the effective resistance is very high. In real junctions, there is usually an increased subgap condutcance due to barrier defects.

ii) A displacement channel through the effective capacitance of the junction. This capacitance is essentially the parallel plate capacitance between the superconductors. This can be rather big as tunnel junctions are typically fabricated in an overlap geometry as discussed in section 5.1.

Nota bene that one can, by combining the two Josephson equations, interpret the small-signal response of the superconducting channel as a nonlinear inductance: Following eq. (2) ϕ is proportional to the time-integral of a voltage and can hence be interpreted as dimensionless magnetic flux, $\phi = 2\pi\Phi/\Phi_0$, where $\Phi_0 = h/2e \simeq 2 \cdot 10^{-15}\text{Vs}$ is the superconducting flux quantum. Thus, we can linearize eq. (1) as $I(\Phi + \delta\Phi) = I_c \sin(2\pi\Phi/\Phi_0) + \delta\Phi/L_J$, or $\delta I = \delta\Phi/L_J(\Phi)$. This defines a kinetic inductance, $L_J(\Phi) = \Phi_0/\left[2\pi I_c \cos(2\pi\Phi/\Phi_0)\right]$ [57,58].

These ingredients can be put together into the famous resistively and capacitively shunted junction (RCSJ) model. It results from Kirchhoff's laws and results in a total current I

$$I_c \sin\phi + \frac{1}{R}\frac{\Phi_0}{2\pi}\dot{\phi} + C\frac{\Phi_0}{2\pi}\ddot{\phi} - I = \delta I(t). \tag{3}$$

where $\delta I(t)$ is current noise. This constitutes the classical equation of motion of a Josephson junction. The ratio of the coefficients in this model can be described by two parameters: The plasma frequency $\Omega_{P0} = (L_J(0)C)^{-1/2} = \sqrt{2\pi I_c/C\Phi_0}$ and the McCumber damping paramter $\beta_c = RC/(L_J/R) = CR^2 I_c/\Phi_0$. The junction is underdamped if $\beta_c \gg 1$. This model has been extensively studied in the classical regime [49,57,58]. It has been derived, in a generalized form, from BCS theory [59].

Aside from these intrinsic elements, one can of course fabricate an artifical shunting circuit on-chip in order to influence the dynamics of the junction.

Already on the classical side, Josephson junction circuits enjoy a huge variety of applications. Superconducting quantum interference devices, SQUIDs [60], are used as ultrasensitive magnetometers and can be used e.g. for measuring brain activity, destruction-free material diagnosis, detection of astrophysical phenomena, high-sensitivity amplifiers, classical flux logic etc. As they are largely controlled by fundamental constants of nature such as e and $\hbar$ and, using eq. (2) can convert frequency into current, Josephson devices also have various applications in metrology [61].

4. Superconducting quantum junctions

In this section, we will derive the macroscopic Hamilton operator of a Josephson junction and the ideas behind it. In the next chapter, we will review several methods to use these junctions in qubits.

Without dissipation as introduced by a shunt conductance (i.e. assuming $R = \infty$), we can rewrite eq. (3) as

$$\frac{\Phi_0}{2\pi} C\ddot{\phi} = I - I_c \sin\phi. \tag{4}$$

This is the equation of motion of a particle with coordinate ϕ and mass $C(\Phi_0/2\pi)^2$ in a tilted washboard potential $U(\phi) = -I\phi\frac{\Phi_0}{2\pi} - E_J \cos\phi$ where $E_J = I_c\Phi_0/2\pi$ is the Josephson energy. We can introduce a Lagrangian

$$L(\phi, \dot{\phi}) = \frac{C}{2}\left(\frac{\Phi_0}{2\pi}\right)^2 \dot{\phi}^2 - U(\phi) \tag{5}$$

whose Euler-Lagrange equation is the correct equation of motion eq. (4). The first term, which plays the role of a kinetic energy, can be interpreted as charging energy, $E_{\rm ch}^{(Q)} = \frac{Q^2}{2C}$, where we have introduced the charge on a capacitor $Q = \frac{\Phi_0}{2\pi}C\dot{\phi} = CV$. It follows, that $\frac{\Phi_0}{2\pi}Q = \frac{\partial L}{\partial \dot{\phi}}$, i.e. it is proportional to the canonical momentum to ϕ. The Hamilton function equivalent to eq. (5) reads

$$H(\phi, Q) = \frac{Q^2}{2C} + U(\phi). \tag{6}$$

So far, we have been treating Q and ϕ as classical variables. Following the canonical quantization procedure, we can readily quantize eq. (6) by identifying ϕ and Q with operators

$$\hat{H} = \frac{\hat{Q}^2}{2C} + U(\hat{\phi}) \qquad \frac{\Phi_0}{2\pi}\left[e^{\pm i\hat{\phi}}, \hat{Q}\right] = \mp\hbar e^{\pm i\hat{\phi}}. \tag{7}$$

This specific form of the commutator would reduce to the usual canonical commutator if ϕ were not a compact variable [62]. For 2π-periodic potentials, this commutator is equivalent to the canonical commuator $\frac{\Phi_0}{2\pi}\left[\hat{\phi}, \hat{Q}\right] = i\hbar$. This is the basis of the macroscopic quantum theory of Josephson junctions. Equation (7) specifically predicts that both ϕ and Q experience quantum fluctuations and cannot be both defined with arbitrary precision; instead, they are limited by a Heisenberg uncertainty relation [40,63]. Typically, the energy scales E_J and $E_{ch} = \frac{2e^2}{C}$ determine the appropriate starting point for describing the junction. For $E_J \gg E_{ch}$ the phase fluctuations are weak, the elementary excitations are quantum vortices, and the charge wildly fluctuates. Junctions of this kind are often termed "classical". In the opposite regime the charge is almost a good quantum number and the phase shows strong fluctuations, these are "quantum" junctions. Using the junction area A, one can estimate $E_J \propto I_c \propto A$ whereas $E_{ch} \propto 1/C \propto 1/A$, thus $E_J/E_{ch} \propto A^2$ and quantum junctions are typically much smaller in area than classical ones. One can show that the charge and the vortex side are dual and one can observe competing order and quantum phase transitions at $E_J \simeq E_{ch}$ [64].

In order to introduce damping and decoherence in a Lagrangian / Hamiltonian formalism, one has to introduce extra degrees of freedom, typically a bath of harmonic oscillators coupling to the junction variables [65,66]. These unobserved degrees of freedom have to be integrated out when making physical predictions for the junction alone. This has been pioneered for a single Josephson junction by Caldeira and Leggett [67,68]. We will now concentrate on junctions with very low intrinsic damping and will describe the remaining dissipation in a way which is compatible with quantum computing in a later section.

For building a quantum bit one has to make sure that single-qubit rotations are possible. This is ensured when the Hamiltonian has off-diagonal terms in the basis of externally controllable variables. Here this means that one has to make sure that *both* variables fluctuate sufficiently, i.e. if a charge-based device is used, one has to provide enough charge fluctuations and vice versa. Several realizations and proposals which accomplish this have been brought forward so far. These range from large single junctions [69–71], highly inductive loops (RF-SQUIDs) [24,72] and small loops [73–76], which are flux-based devices to Cooper pair boxes [30,77–80], which are charge-based. Other devices combine charge and phase fluctuations with comparable strength and operate in between [81]. Other proposals use the specific properties of unconventional superconductors [82,83] or large arrays [84].

5. Different types of superconducting qubits

In this section, we will describe different classes of the Josephson quantum bit along with their experimental status. Superconducting qubits can be devided into roughly three groups: "charge", "phase", and "flux" qubits.

As mentioned in the previous sections, the common essential element of the superconducting qubit is the mesoscopic scale Josephson junction. In order to prepare a qubit, we need well defined quantum two state system whose intrinsic energy scales are lower than the superconducting energy gap and to a high degree isolated from the outside world. The relevant two energy scale to describe a junction are the Josephson energy E_J and the charging energy E_{ch} of the junction for a single Cooper pairs. In the case of junctions made from aluminum thin films, which is the most common technology in the field, these energy scales become comparable typically in the regime of submicron sized junctions. So the superconducting Josephson qubit is basically a nonlinear electronic circuit made of superconductor with submicron width and thickness, which we can fabricate by the modern electon beam (EB) lithography and shadow evaporation technique[85].

5.1. *Sample preparation*

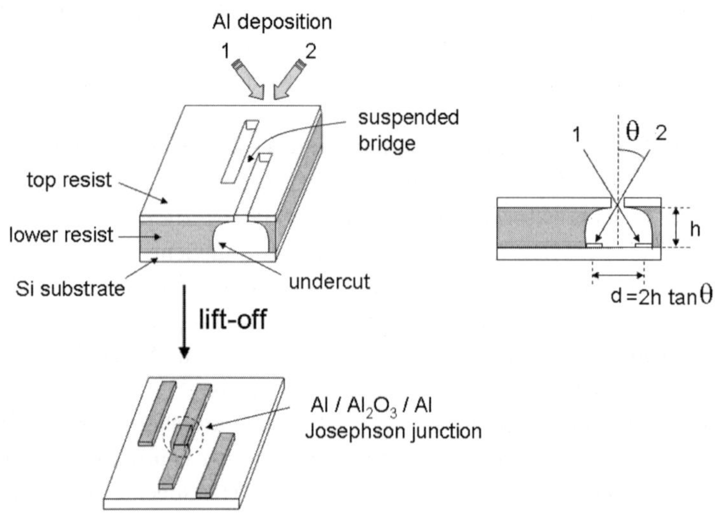

Fig. 1. Illustration of suspended bridge and shadow evapolation technique .

Before mentioning the different qubit design, here, we will briefly mention the most popular process to fabricate the aluminum Josephson junction. First, one has to prepare the double layered resist structure with two different kind of polymers on the thermally oxidized silicon substrate. As illustrated in Fig. 1, the upper thin layer is made of less EB sensitive posi-typed resist. The lower spacer layer is made of more EB sensitive posi-typed resist. After EB pattern writing and developping process with chemical developper, we get a well definied bridge pattern with a nice undercut in the lower layer. Then, the first few tens of nm thick aluminum layer is evaporated from an angle $-\theta$ from the vertical line. Next, the sample is exposed to dilute oxigen gas. We expect the very thin natural alumina tunnel barrier layer to cover the surface of the fresh aluminum thin film. After purging the oxygen gas, the top aluminum layer is evaporated from the angle of θ. Assuming the thickness of the lower spacer layer is h, the distance between two pieces of evapolated aluminum pattern is $d = 2h \tan\theta$. By designing h and θ in advance, we get submicron scale $Al/Al_2O_3/Al$ Josephson junction through the lift-off process. This Dolan method[85] has been widely used in reliably making small Josephson junctions over a quarter century.

5.2. *The flux qubit*

A paradigmatic type of Josephson qubit is a flux qubit. It descends from the mesoscopic scale rf-SQUID first introduced by A. J. Leggett to discuss macroscopic quantum tunneling and macroscopic quantum coherence in his groundbreaking paper[23] a quarter century ago.

We will now describe the contemporary version of this device in more detail, namely the persistent current quantum bit, which is phase-based and uses a small loop. This device has been proposed at TU Delft and MIT[73]. As all approaches it has its specific strengths and weaknesses. Most of the theoretical work compiled in the later sections are motivated by this device, but the main ideas can be described in terms of universal Hamiltonians and can be applied to other setups as well.

The flux qubit, fig. 2, consists of a micrometer-sized superconducting loop, which is interrupted by three Josephson tunnel junctions made from the same, conventional technology: Two of equal size, one smaller by a factor $\alpha \simeq 0.8$. The loop dimensions are chosen such that the geometric self-inductance of the loop does not play any significant role. The loop is penetrated by a magnetic flux of size Φ_x, which imposes the quantization condition $\phi_1 + \phi_2 + \phi_3 = f$, where $f = (2\pi\Phi/\Phi_0) \bmod 2\pi$ is the magnetic frustration, for the phases across the three junctions. Thus we can eliminate

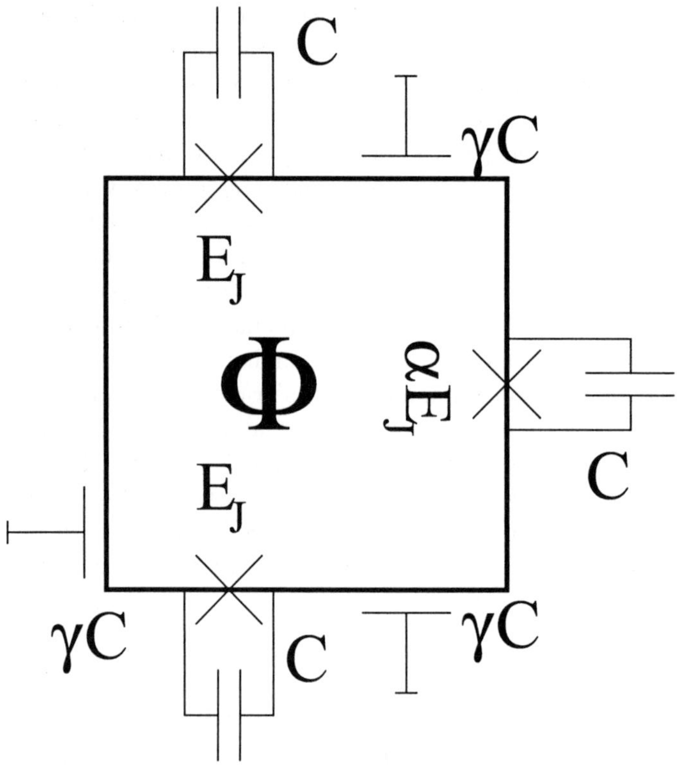

Fig. 2. The circuit diagram of a flux qubit threaded by a magnetic flux Φ, including three Josephson junctions (crosses) and all geometric and stray capacitances .

ϕ_3, the phase across the weaker junction, and obtain the potential energy

$$U = E_{\mathrm{J}}\left(-\cos\phi_1 - \cos\phi_2 - \alpha\cos(f + \phi_1 - \phi_2)\right). \tag{8}$$

This potential is plotted in figure 4. The potential is periodic and possesses a hexagonal pattern of minima separated by potential wells. The energy difference of adjacent minima can be tuned through the external flux: They are degenerate at $\Phi = \Phi_0/2$. From choosing one of the Josephson junctions smaller than the others, one direction is introduced in which the potential barrier is substantially smaller than in the other directions. The state in the minima correspond to clockwise and counterclockwise circulating current respectively.

The charging energy can be evaluated from Kirchhoff's laws. The result is written using vectors in the two dimensional $\{\phi_1, \phi_2\}$ and $\{Q_1, Q_2\}$

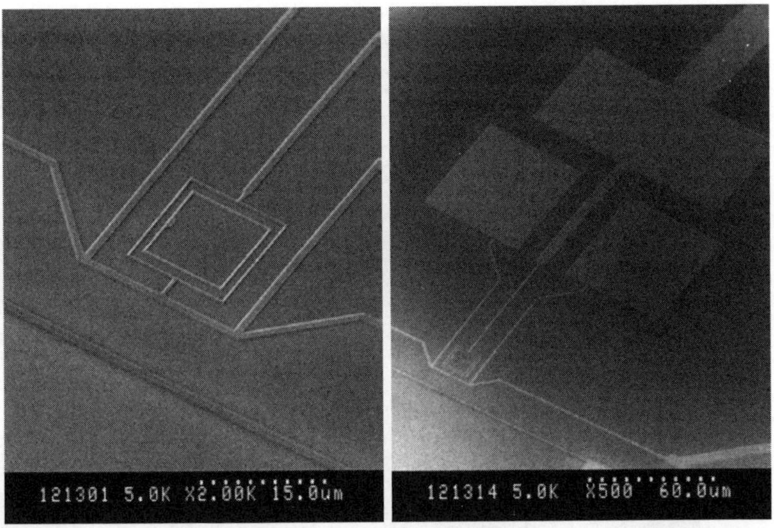

Fig. 3. SEM photographs of a flux qubit fabricated at NTT-BRL Atsugi Labs. The aluminium thin film sample (thickness ~0.1μm) is fabricated on a thermally oxidized silicon substrate with standard electron beam lithography and subsequent shadow evaporation of aluminium. The qubit is the inner square loop enclosed by the quantum detector SQUID which are clearly seen in the left close-up view. In the right photo, the square plates at the top of the picture are the top plates of the on-chip capacitors separated by an aluminium oxide insulator layer from the larger bottom plate. The qubit is controlled through the oscillating magnetic flux produced by an ac-current through the on-chip microwave line which is 15 μm away from the qubit.

coordinate-space as

$$E_{\text{kin}} = 2e^2 \vec{Q}^T \mathbf{C}_{\text{M}}{}^{-1} \vec{Q}. \tag{9}$$

with a capacitance matrix

$$\mathbf{C}_{\text{M}} = C \begin{pmatrix} 1+\alpha+\gamma & -\alpha \\ -\alpha & 1+\alpha+\gamma \end{pmatrix}. \tag{10}$$

Here, γ is the ratio of the stray capacitances to ground over the junction capacitances, as seen in figure 2.

Carefully choosing appropriate parameters, one can reach a situation with exactly one bound state per minimum, where the tunneling amplitude along the easy direction is substantial and is strongly suppressed along the other directions. At low energies the dynamics of the system can be described in a two-state approximation in the basis of the states localized

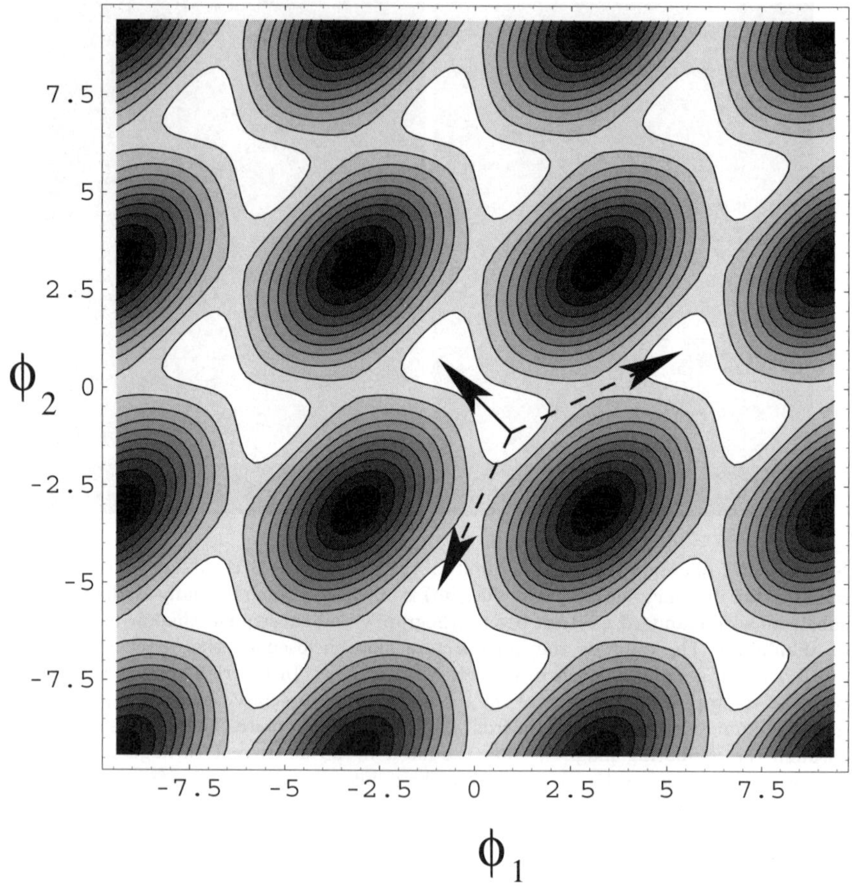

Fig. 4. The potential landscape for a flux qubit as a function of the phases across the identical junctions, taking $\alpha = 0.8$. The solid arrow indicates an easy tunneling path whereas the dashed lines indicate hard directions.

in the potential minima, the classical states,

$$\hat{H}_2 = \frac{1}{2}\begin{pmatrix} \epsilon & \Delta \\ \Delta & -\epsilon \end{pmatrix}. \tag{11}$$

The energy bias ϵ can be steered through the external flux following $\epsilon \simeq I_{\mathrm{q}}(\Phi - \Phi_0/2)$, where I_{q} is the modulus of the circulating current in the classical states. Thus, the quantum dynamics of the device can be controlled by the external flux.

The tunnel splitting Δ can be made tunable by splitting the small junction into two parallel junctions, a DC-SQUID, which acts as an effective single junction whose Josephson coupling $E_J(\Phi_2) = 2E_{J,0}\cos(\pi\Phi_2/\Phi_0)$ can be tuned by the flux Φ_2 through this loop between the sum of the two couplings $2E_{J,0}$ and zero. Such a tunable Δ is desirable in a number of, but not in all, quantum computing protocols. The state of the system can be read out by measuring the extra magnetic flux produced by the circulating current through a very sensitive magnetometer: a SQUID[60]. Such a magnetometer works as a tunable junction as just described: E_J depends on the flux through the loop and can be measured electronically. The SQUID-readout corresponds to a measurement of $\hat{\sigma}_z$. Figure 3 shows micrographs of real devices together with their read-out apparatus.

The flux qubit is thus a well-defined quantum system that can perform single-qubit rotations. All other ingredients demanded by DiVincenzo's original five criteria can also be met, which will be detailed more later in this chapter. Note that eq. (11) predicts that superpositions of current states can be prepared close to $f = \pi/2$. The current states involve up to 10^{10} electrons. Thus, these are superpositions of large objects. This does not yet imply that these states correspond to huge Schrödinger's cats: For analyzing this question one has to carefully evaluate the distance in Hilbert space between the two states, which is a by far more subtle issue [86,87].

Note that the analysis here for simplicity neglected self-inductancs, which can be included and does not change the overall picture [88,89]

In the advanced flux qubit layout shown in Figure 3 there are two important control parameters of the circuit : the externally applied bias magnetic flux Φ_{ext} and the bias current through the SQUID I_{b}. In the flux qubit illustrated in Fig. 5, qubit and SQUID share two edges in order to use the kinetic inductance (the kinetic energy of the Cooper pairs in a line, similar to the kinetic inductance of a junction, see section 3) to increase the coupling strength. This also automatically enables to adiabatically shift the flux bias during the I_{b} pulse for qubit state readout just after operating qubit at the *optimal* point. Taking into account that this device is made from a double layer structure and there are odd number junctions in a qubit loop, the persistent currents in the sharing edges do not have to flow in the same layer, whereas the bias currents I_{b} through the SQUID flow in the same lower layer. Unfortunately, this fact causes considerable asymmertric coupling due to kinetic inductance of the qubit and SQUID[90]. Thus the current noise via I_{b} through the SQUID can be dominant even at optimum flux bias point where $\frac{dE_{01}}{d\Phi_{\text{ext}}} = 0$. In general, this type of asymmetry always exists even in the system of original type flux qubit surrounded by a SQUID as shown in Fig.3 top if there is difference in critical current of two Josephson junctions in the SQUID.

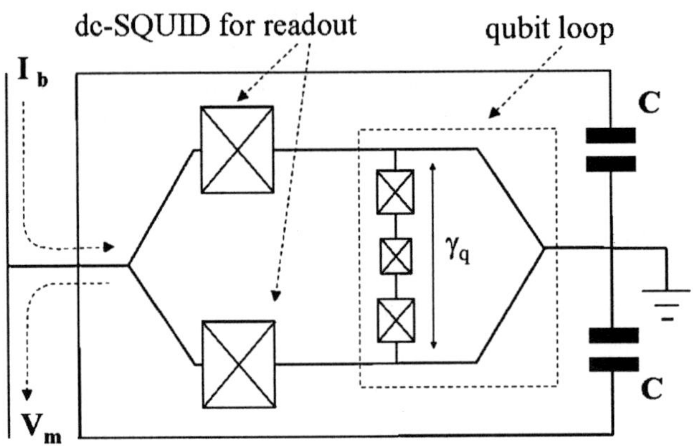

Fig. 5. Schematic circuit diagram of the advanced version of the Delft type flux qubit. The qubit loop and the SQUID are now galvanically connected. The SQUID is shunted by additional capactiors C. I_b is the bias current and V_m the detection voltage.

To overcome these difficulties, the two complementary strategies to protect the qubit from these decoherence sources are employed. One consists in biasing the qubit so that its resonance frequency is stationary with respect to the control parameters (*optimal point*), see section 7.1 ; the second consists in *decoupling* the qubit from current noise by choosing a proper bias current through the SQUID, see section 7.3. Using these strategies, finally at the decoupled optimal point, long spin-echo decay times of up to 4 μs were successfully demonstrated[91].

5.3. *Other designs*

Let's have a look at different schemes of superconducting qubits, one by one. First, as illustrated in Fig. 6, the essential component of the "charge" qubit is a Cooper pair box, a submicron scale superconducting island which is connected to the ground via small Josephson junction. Here, the box is in the charging energy dominant regime $E_{\rm ch} > E_{\rm J}$. In order to tune the $\frac{E_J}{E_{\rm ch}}$ ratio of the system, this Josephson junction can be separated into two parallel junctions forming a SQUID then one can manipulate $E_{\rm J}$ value by controlling the magnetic flux $\Phi_{\rm ext}$ which penetrates the SQUID loop. In addition to the Josephson junctions, the qubit is also equipped with two lines.

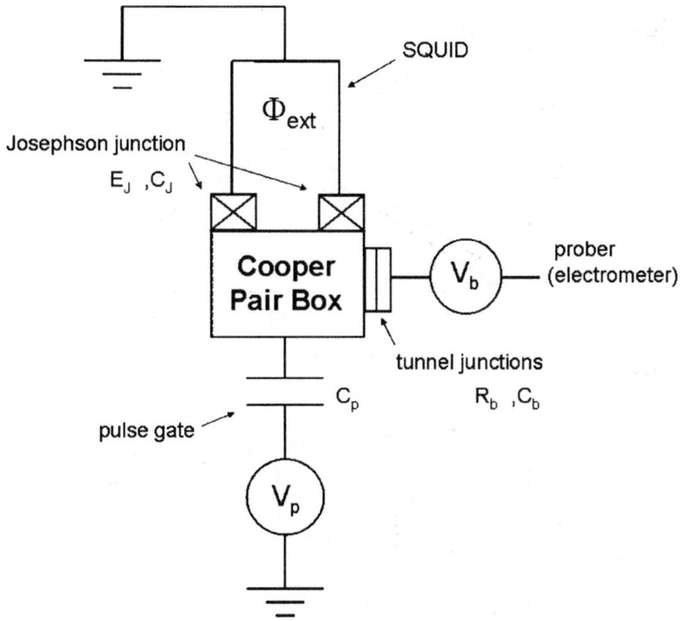

Fig. 6. Josephson charge qubit. The central component is a tiny superconducting island called Cooper pair box equipped with small Josephson junction(s) where the charging energy dominates the Josephson energy i.e., $E_{ch} > E_J$. This E_J value can be tunable by changing the externally applied magnetic flux Φ_{ext} in the SQUID loop. The qubit energy is a function of both gate induced charge n_g and Φ_{ext}. The qubit state is controlled by changing n_g or applying resonant microwave pulse from the onchip RF-line .

One is a probe electrode as a measurement readout line which is weakly tunnel coupled to the Cooper pair box through a thicker oxide layer shown as a tunnel junction in Fig. 6. The other one is a control pulse gate which is spatially separated from the Cooper pair box and only capacitively coupled to it. In the charge qubit, a good quantum number is a number of excess Cooper pairs $|n\rangle$ in the box. If one can control the gate voltage precisely enough, then one can change the induced charge n_g by the gate voltage. The the two states $|0\rangle$ and $|1\rangle$ implement a qubit in this system: they would be classically degenerate when $n_g = 0.5$ but are coupled through E_J and however, the Josephson tunneling energy E_J will lift the degeneracy illustrated in Fig. 7(a). The NEC group prepared charge state $|0\rangle$ at $n_g = 0.5$ as shown in Fig. 7(b). Using a non-adiabatic (very fast on the scale of the qubit level splitting) gate pulse with rise and fall time of 30$\sim$40 ps and Δt = 80 ps $\sim$ 2 ns duration, they then succeeded in observing internal

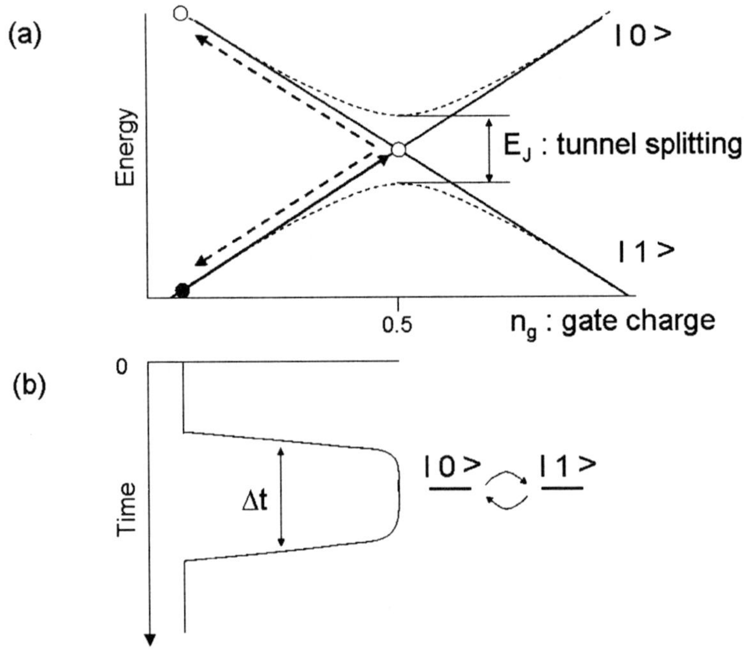

Fig. 7. Energy diagram (a) and the control scheme (b) of the Josephson charge qubit. The qubit is controlled via the gate induced charge n_g in the Cooper pair box by applying a fast (non-adiabatic) voltage pulse from the pulse gate. (a) Initially, qubit state is $|0\rangle$ (shown as a black dot). Next, bring the qubit state to the degeneracy point ($n_g = 0.5$) by the non-adiabatic gate pulse whose rise and fall time is shorter than the time scale determined by the interaction energy $\frac{h}{E_J}$. (b) Qubit state starts quantum oscillation between $|0\rangle$ and $|1\rangle$ during the time period of Δt.

coherent oscillation between two energy eigenstates[77] which are shown as the dotted curves in the Fig. 7(a). With an array of pulses with a repetition time longer than the relaxation time, one can repeat the pulse operation many times and measure the direct current through the probe junction which would reflect the population in $|1\rangle$ after each pulse operation. The experiment demonstrated quantum oscillation with a period of $\frac{h}{E_J}$ up to 2 ns. This was the first example of the long sought evidence of macroscopic quantum coherence[92]. Other than the gate pulse operation, this type of qubit can be also operated by the resonant microwave pulse[93].

It is worth mentioning that the echo technique described in section 7.6 works very effectively in the charge qubit. In fact, the coherence time during free evolution T_2^* was markedly improved more than an order of magnitude[78] i.e., from 0.1 ns to the order of nanoseconds.

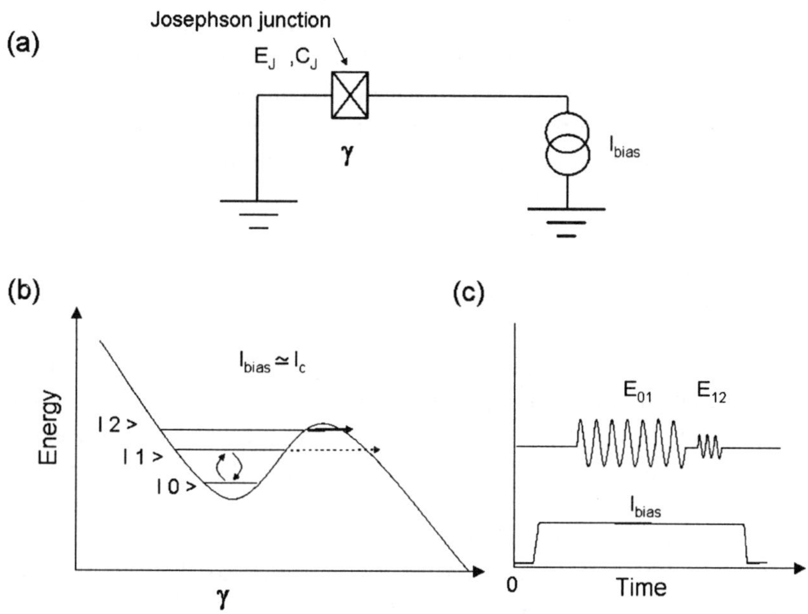

Fig. 8. Josephson phase qubit (a) circuit diagram (b) energy diagram (c) timing of the bias current pulse (I_{bias}), microwave pulses for qubit operation (indicated as E_{01}) and readout (E_{12}).

This implies the main origin of dephasing in this charge qubit can be ascribed to the low frequency charge noise generated by the random motion of background offset charges, see section 7 of this chapter.

A further type of qubit is the phase qubit which is illustrated in Fig. 8(a). The Josephson phase qubit is a single Josephson junction with an extremely precise current bias. In this layout, the lowest two quantum levels in a washboard potential shown in Fig. 8(b) are used to implement a qubit. The quantum coherent oscillation of the phase qubit were first observed by Martinis[69] and Yu[70] independently. The quantum properties of current biased Josephson junction are well established[94]. As long as the dc voltage across the junction is zero, the nonlinear Josephson inductance and the junction capacitance form an anharmonic LC-resonator. The two lowest quantized energy levels are the states of the qubit. By design the energy separation E_{01} of these lowest two states can be set much larger than the operating temperature $E_{01} \sim 10k_{\rm B}T$. The sequence of initialization, operation and readout of the qubit is schematically illustrated in Fig. 8(c). In order to keep only a few states in an approximate cubic potential, the current bias $I_{\rm bias}$ is pulsed for a time $\sim 50\ \mu$s, which is typically driven close to the critical current $I_{\rm c}$ of the junction. This is an initialization procedure. Then, a microwave pulse resonant with E_{01} is applied to induce Rabi oscillations. Immediately after the microwave pulse for Rabi oscillation between ground state $|0\rangle$ and first excited state $|1\rangle$, the readout microwave pulse resonant with E_{12} is applied to the qubit. If and only if the qubit was in state $|1\rangle$ after the first Rabi pulse, it will proceed to state $|2\rangle$. As state $|2\rangle$ is close to the top of the barrier it easily escapes to the running state and leads to a finite voltage which can be detected. The huge difference in the escape rate between excited levels and the ground state makes it possible to use this as a detector without destabilizing states $|0\rangle$ and $|1\rangle$. Repeating this measurement sequence typically $10^4 \sim 10^5$ times, one can obtain information about the qubit state through occupation probability of the excited state. As a phase qubit, a rather large ($\sim$10 μm $\times$10 μm) current biased Nb-based Josephson junction has been used. The most significant advantage of the phase qubit is scaling to more complex circuits will be favorable because fabrication and operation of complex superconducting integrated circuit with *large* junctions are well established.

Here, we would like to mention about the quantronium project of the Saclay Quantronics group[81], a hybrid between the two types described so far. The quantronium circuit is an Al-made Cooper pair box equipped with a large Josephson junction with Josephson energy $E_{\rm J0} \approx 20E_{\rm J}$ as a readout device. This qubit has two control knobs, the gate induced charge $n_{\rm g}$ in the Cooper pair box and the bias flux $\Phi_{\rm ext}$ imposed through the loop. The remarkable strategy of this project is to protect qubit from the noise which comes from the electro-magnetic environment by using an *optimal* operating point such as a saddle point or an extremum where $\frac{\partial E_{01}}{\partial n_{\rm g}} = 0$ and $\frac{\partial E_{01}}{\partial \Phi_{\rm ext}} = 0$. Moreover, the qubit has a large energy splitting $\frac{\Delta}{h} \sim 16$ GHz which means rather flat energy dispersion and also intermediate-energy regime i.e., $\frac{E_{\rm J}}{E_{\rm ch}} \sim 1$ where a circuit is designed to be insensitive to fluctuation of both charge and flux bias. Thanks to the above mentioned ingenuities, they achieved a successful outcome i.e., the Ramsey measurement shows the decay time of $T_2^* \simeq 0.5\mu$s during free evolution by operating the quantronium at the saddle point of the control parameters.

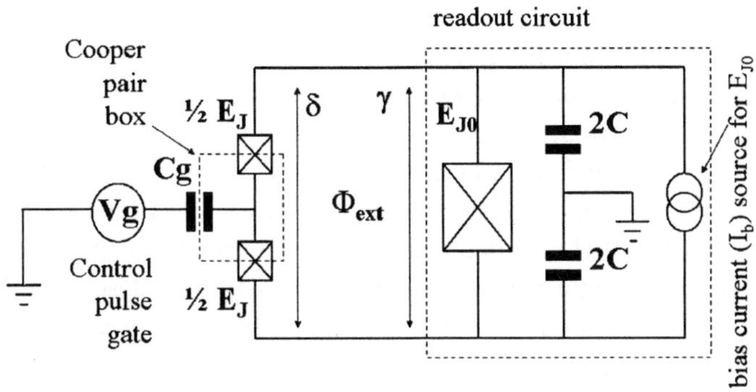

Fig. 9. Schematic circuit diagram of the quantronium. A hybrid type design of a charge-qubit (a Cooper-pair box) and a large Josephson junction as a readout circuit which is essentially the same as a phase-qubit. By way of the uniqueness of the phase in a superconducting loop which contains both the Cooper-pair box and the large Josephson junction, the phase difference δ across the Cooper pair box, is entangled with the phase difference γ across the large junction which is related with two quantum states of the current biased large Josephson junction. Thus, the state of the charge-qubit is readout through the state of the phase-qubit. This qubit has two control knobs, the gate induced charge n_g in the Cooper pair box and the bias flux Φ_{ext} imposed through the loop. The qubit energy is a function of both n_g and Φ_{ext}. By operating the quantronium at the double extremum condition, they succeeded to obtain remarkablly long coherence time during free induction decay $T_2^* \simeq 0.5\mu s$.

6. Experimental achievements

In this section, we are going to overview the status of experimental forefront research of different kind of Josephson qubits.

6.1. *Single qubit operation*

For the single qubit rotations, the resonant microwave driven Rabi oscillations is commonly used[69,70,75,81,93]. On the other hand, the Larmor precession induced by the non-adiabatic fast pulse can be also used[77]. Rabi oscillations with a characteristic decay time of a few microseconds were observed in the phase qubit where current biased large Nb-based Josephson junction (typically $\sim$10 μm $\times$10 μm) was used[70].

In quantum computation, it is essential to control each qubit by performing arbitrary unitary operations at will. For one qubit, Rabi oscillation and Ramsey fringes experiments provide information related to the control of the qubit state $|\Psi\rangle = \cos\frac{\theta}{2}|0\rangle + e^{i\phi}\sin\frac{\theta}{2}|1\rangle$. In order to achieve noise torelant qubit operation, NMR-like multi pulse sequence control has been demonstrated in Josephson charge-phase qubit by the Saclay group [95]. However, the observation of the Ramsey fringes of a flux-qubit usually involves a few hundred MHz detuning from the qubit resonant frequency.

The NTT group proposed a new method[96] for observing Ramsey fringes, the phase shift method, which can control the phase of microwave (MW) pulses at the resonant frequency of the qubit. The advantage of this method is that it provides qubit rotation along an arbitrary axis in the x-y plane and the faster control of the azimuth angle ϕ of a qubit than the conventional detuning method. Figure 10 shows schematic diagrams describing how the qubit vector is operated during the Ramsey fringe experiment with the phase shift technique in a rotating frame.

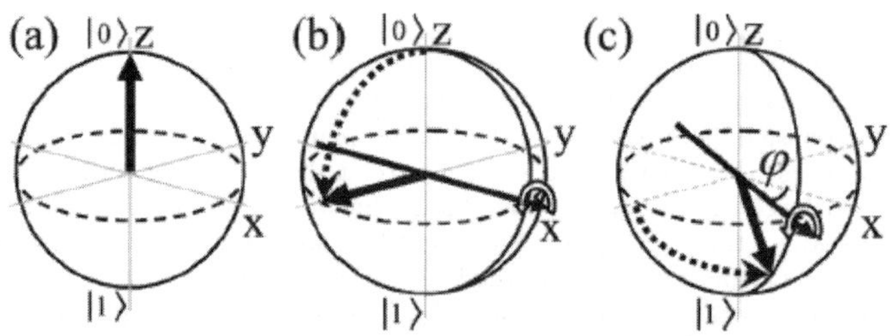

Fig. 10. Schematic diagram of qubit vector motion induced by the phase-shifted double $\frac{\pi}{2}$ on-resonance pulses ($\omega = \omega_0$). It is described in the rotating frame of the qubit Larmor frequency ω_0. (a) The qubit vector in the initial state. The qubit is in the ground state $|0\rangle$. (b) The first resonant $\frac{\pi}{2}$ pulse ($\varphi = 0$) tips the qubit vector to the equator. The qubit vector remains there, because the on-resonance pulse is used. (c) The second resonant $\frac{\pi}{2}$ pulse, in which the phase-shift $\varphi \neq 0$ is introduced, tips the qubit vector on another axis, which is at an angle φ from the x-axis. Reused with permission from Appl. Phys. Lett. **87**, 073501 (2005). ©2005, American Institute of Physics.

We assume that the initial state of the qubit is the ground state $|0\rangle$. The first resonant $\frac{\pi}{2}$ pulse ($\varphi = 0$) tips the qubit vector towards the equator with the x-axis as the rotating axis ($\hat{H}_{\text{rot}} \propto \hat{\sigma}_x$). The qubit vector remains there because we introduce no detuning ($\omega = \omega_0$). After a time t_{12}, the second resonant $\frac{\pi}{2}$ pulse with a given phase shift $\varphi \neq 0$ tips the qubit vector on another axis at an angle φ from the x-axis. The resulting qubit vector does not reach the south pole ($|1\rangle$) of the Bloch sphere. The detector SQUID switches by picking up the z-component of the final qubit vector after the trigger readout pulse. Repeating this sequence typically 10,000 times, with a fixed t_{12}, we obtain the switching probability. Figure 11 shows the damped sinusoidal oscillation obtained by changing the pulse interval t_{12}. The phase shift of the second pulse was programmed from the following relation ; $\varphi = \omega_0 t_{12} \bmod 2\pi$. This equation gives a 2π phase change to the resonant microwave pulse during a period of $T = \frac{2\pi}{\omega_0}$. This means that we introduce a phase shift with the Larmor frequency.

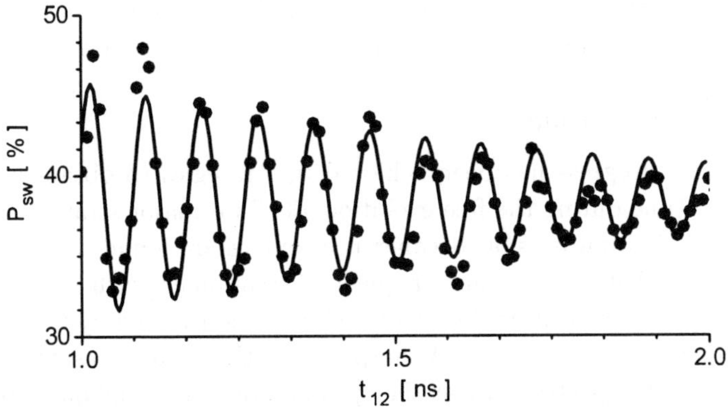

Fig. 11. On-resonance Ramsey fringes observed by using the phase-shifted double $\frac{\pi}{2}$ pulse technique. The Larmor frequency is $\omega/2\pi$=11.4 GHz. The width of the $\frac{\pi}{2}$ pulse, 5 ns, is determined by Rabi oscillation. An exponentially damped sinusoidal curve fitted with the decay time constant T_2 = 0.84 ns is also shown. Reused with permission from Appl. Phys. Lett. **87**, 073501 (2005). ©2005, American Institute of Physics.

When Ramsey fringes are observed in the conventional way, a few hundred MHz detuning is typically introduced near the qubit Larmor frequency, i.e., $\sim$ 100 MHz detuning at a Larmor frequency of $\sim$ 5 GHz. With this detuning method, after the first detuned $\frac{\pi}{2}$ pulse, the qubit vector rotates along the equator of the Bloch sphere with this detuning frequency, $\sim$ 100 MHz [75]. If we use this method to control the qubit azimuth angle, a time of $\sim$ 10 ns is required for every 2π azimuth angle rotation of the qubit vector. This operating time cannot be as short as 1 ns, because a detuning of 1 GHz does not work properly. However, with the phase shift technique with the resonant frequency, as we have shown, it is possible to revolve the rotational axis of the qubit vector within the xy-plane with the frequency above 11 GHz. By way of the well known decomposition of a rotation in three rotations around two orthogonal axes[4], one can perform quick qubit rotations without using relatively slow free evolution by introducing detuning. In particular, using the relation $Z(\phi) = X(\frac{\pi}{2})Y(\phi)X(-\frac{\pi}{2})$, the azimuth angle ϕ rotation on the z-axis can be decomposed into three successive rotational operations such that $-\frac{\pi}{2}$ rotation on the x-axis, ϕ rotation on the y-axis, and $\frac{\pi}{2}$ rotation on the x-axis. If the qubit is driven strongly enough, each $\frac{\pi}{2}$-pulse width can be as short as 0.1 ns, therefore the total composite operation $X(\frac{\pi}{2})Y(\phi)X(-\frac{\pi}{2})$ can be completed in $\sim$ 1 ns. Compared with the conventional detuning method, the phase shift technique provides us with the opportunity to increase the speed of the qubit unitary gate operation by more than an order of magnitude. This method will save operating time and we can make best use of the precious coherence time.

6.2. *Coherence time*

Remarkable progress has been achieved in strategies to obtain improved coherence time during the free evolution T_2^*. Two major strategies to protect the qubit from possible decoherence sources are found. One is biasing the qubit so that its resonance frequency is stationary with respect to the control parameters (*optimal point*) which is demonstrated by Saclay group in the "quantronium" project. A sharp line width as narrow as 0.8 MHz was observed in the spectroscopy at the optimal point with the qubit resonant frequency $\sim$ 16GHz, where a Q-value over 2×10^4 was obtained. Just like the pulsed NMR technique, they obtained an optimised $T_2^* \simeq 0.5\mu$s of their charge-phase qubit from a Ramsey fringe experiment[81]. Now, this strategy is already a common standard in most experimental efforts. The second strategy was applied in the flux-qubit experiment, which consists in *decoupling* the qubit from current noise by choosing a proper bias current through the readout SQUID. Using 4- instead of 3-junction qubit, at the decoupled optimal point, the long spin-echo decay times of 4 μs was demonstrated[91]. As these strategies decouple from different sources of T_2 limitation, they conceivably need to be combined to achieve further progress.

In usual experiment, we observe T_2, the transversal relaxation or dephasing time, is (much) shorter than the energy relaxation time T_1. From a general consideration, the energy relaxation time T_1 is governed by the noise spectral density $S(\omega)$ at a frequency corresponding to the qubit level splitting, as described by Eq. (12) in chapter 7.1. On the other hand, the phase relaxation time T_2 is governed by the noise spectral density at low frequency limit $S(0)$ but also contains T_1 Eq. (13). In the idealistic situation where $S(0) \to 0$, the relation $T_2 = 2T_1$ is expected. Recently, JST-RIKEN-NEC group[97] and NTT group[98] have observed the $T_2 \approx 2T_1$ relation at the optimal operating point.

In the phase qubit, large-gap superconductors such as Nb or NbN have an advantage in supressing unwanted quasi-particle generation due to thermal activation or due to parasitic circuit resonances inherently excited in the on-chip circuit. However, a lot of anticrossings in the microwave spectra[99] and the observation of real-time oscillations between qubit and resonator [100] reported by the NIST group provide evidence that some kind of unexpected fluctuators are coupled coherently to the phase qubit. They observed that a number of small such spurious resonators have a distribution in splitting size, with largest one giving a splitting of $\sim$ 25 MHz and an approximate density of 1 major spurious resonance per $\sim$ 60 MHz. When thermally cycled from $\sim$20 mK to room temperature, the magnitude and frequency of the spurious resonances changed considerably, whereas cycling to 4 K produces no apparent effect. Furthermore, over 10 qubit devices with the same experimental setup, they found that each qubit has its own unique "fingerprint" of resonance frequencies and splitting strength. From these observations, they ascribed that these spurious resonances are microscopic in origin such as two-level fluctuators within the amorphous oxide tunnel barrier in the Josephson junction which couple to the qubit state through the critical current. This should be one of the origin of similar magnitude of $1/f$ noise observed over the tunnel junction made from oxide of Al, Nb, PbIn. In that work, it is insisted that improvements in the coherence of all Josephson qubits will require materials research directed at reducing or removing these resonance states that have remained hidden for over 40 years in the conventional junction fabrication technique such as shadow evaporation. Following this warning, the NIST and also the Delft group already started projects to fabricate Josephson junction with greater microscopic uniformity by employing molecular beam epitaxy.

6.3. *Different materials*

Here, we mention about two more topics from the material related side of qubit research. Recently, two groups independently succeeded in observation of macroscopic quantum tunneling (MQT) in Josephson junction made of d-wave high-Tc cuprate superconductors. The Chalmers group used YBCO grain boundary biepitaxial junction[101]. The Tohoku university group, using an intrinsic Josephson junction in BSCCO, observed a quantum-classical crossover temperature for MQT of $T_{MQT} \sim 1$ K [102] which is significantly larger compared with $T_{MQT} \leq 300$ mK of conventional superconducting material such as Al or Nb. This result indicate the merit of using a superconductor with a larger gap energy scale exceeds the demerit that there are gapless excitations in some directions on the Fermi surface. The other topic is that all MgB_2 tunnel junctions showed Josephson current-voltage characteristics above 20 K[103,104]. The binary compound MgB_2 has $T_c \sim 40$ K would be another candidate to fabricate noise tolerant qubit which requires a large energy splitting within the superconducting gap.

6.4. *Readout*

A good readout visibility is another indispensable condition for quantum computation. The NTT group reported an interesting possibility[105]. They used a detection SQUID with very small Josephson junctions (0.1 μm $\times$ 0.1 μm). The qubit signals were measured in a regime of small switching current I_{sw} of the SQUID, typically less than 100 nA, where the switching current distribution turns out to be particularly narrow. The obtained single-shot data indicate that the qubit state is readout as energy eigenstates rather than current eigenstates although I_{sw} of the SQUID were used as a detection observable. In this case, I_{bias} of the SQUID were swept very slowly. The electro-magnetic environment of SQUID was likely to be substantially overdamped. They interpreted their result as follows[106] : the qubit-SQUID interaction energy was so small that the qubit energy eigenstate was readout without being projected onto the current states. In other words, while qubit state is bimodal, however, the SQUID switching current I_{sw} can take a continuous value. If the qubit energy splitting is larger compared with the qubit-SQUID interaction energy, the SQUID only switches under the influence of magnetic flux created by the qubit.

The RIKEN-NEC group reported that they succeeded in single shot state readout of the Josephson charge qubit by using an onchip superconducting single electron transister (S-SET)[107]. The quantum bits were transformed into and stored as classical bits (charge quanta) in a dynamic memory cell - a superconducting island. The transformation of state $|1\rangle$ (differing from state $|0\rangle$ by an extra Cooper pair) was a result of a controllable quasiparticle tunneling to the island. The charge was then detected by a conventional single-electron transistor, electrostatically decoupled from the qubit. They also studied relaxation dynamics in the system and obtained the readout efficiency of 87 and 93% for $|1\rangle$ and $|0\rangle$ states, respectively.

Recently, it has become clear that in the SQUID switching detection type measurement, we lose substantial qubit visibility during relatively slow ramp up process of the SQUID bias current. In fact, by changing the conventional bias readout pulse of 10 ns rise time to faster microwave pulse of 0.5 ns rise time by the resonant activation method, 40% visibility of the flux-qubit was improved up to 65% [108].

On the other hand, another approaches of qubit state measurement without switching the SQUID detector are in progress in Yale[109,110] and also in Delft[111]. On the superconducting branch, the SQUID acts as if it is an inductance (Josephson inductance), see section 3. They use the circuit which contains SQUID junctions and the onchip capacitor as an anharmonic LC-resonator and detect resonant frequency shift as a function of applied magnetic flux. The intrinsic flux detection efficiency and backaction are suitable for a fast and nondestructive determination of the quantum state of the qubit. This method will provide an opportunity of quantum non-demolition measurement of a qubit as needed for readout of multiple qubits in a quantum computer.

6.5. *Two qubit operation*

After the initial breakthrough in the coherent manipulation of a single Josephson qubit[77], the next target has been to control entanglement in two-qubit system. The NEC-RIKEN group has reported a controlled-NOT operation in the capacitively coupled two Josephson charge qubits[112]. They could apply fast (rise/fall time 40 ps) voltage pulse to change each qubit state independently. For those input states ($|10\rangle$ and $|11\rangle$ the target qubit should not be flipped), with the control qubit state was $|1\rangle$, their gate operation was almost ideal. On the other hand, when the control qubit state was $|0\rangle$ (input states of $|00\rangle$ and $|01\rangle$, the target qubit should be flipped), the output states contained an unwanted target qubit component ($|00\rangle$ or $|01\rangle$) with a rather high probability ($\sim$37%). According to them, this might be due to the finite rise/fall time (40 ps) of the operation pulse, which increased the unwanted oscillation.

Recently, the NIST-UCSB group has succeeded in simultaneous state measurement of capacitively coupled Josephson phase qubits[113]. They have used simultaneous single-shot measurement of coupled Josephson phase qubits to directly probe interaction of the qubits in the time domain. The concept of measurement crosstalk is introduced, and they showed that its effects are minimized by careful adjustment of the timing of the measurements. They observed an antiphase oscillation of the two-qubit $|01\rangle$ and $|10\rangle$ states.

6.6. *Cavity QED on a chip*

The alternative way to implement multi qubit gate is using a quantum bus in order to couple arbitrary pair of qubits on demand. This Cirac-Zoller type scheme[13] is used in ion trap systems. Recently, observation of strong coupling in all- solid state implementations of the cavity QED concept using circuit elements were reported by the Delft-NTT-NEC group and the Yale group independently. Delft-NTT-NEC group[114] has observed a flux qubit strongly coupled to non-linear SQUID oscillator. On the other hand, the Yale group[115] has observed a charge-phase qubit strongly coupled to a dispersive transmission line resonator, which acts as a linear oscillator.

The Yale group also succeeded in observing the ac Stark shift[116] of the qubit level in a single photon resolution. This means, a single photon on average and does not refer to a single photon Fock state. They have demonstrated that the strong coupling of a Cooper pair box to a non-resonant microwave field in an on-chip cavity gives rise to a large qubit dependent shift in the excitation energy of the resonator. As the counteraction, the ac Stark effect also shifts the qubit level separation by about one linewidth per photon at 2% detuning, and the backaction of the fluctuations in the field gives rise to a large broadening of the qubit line.

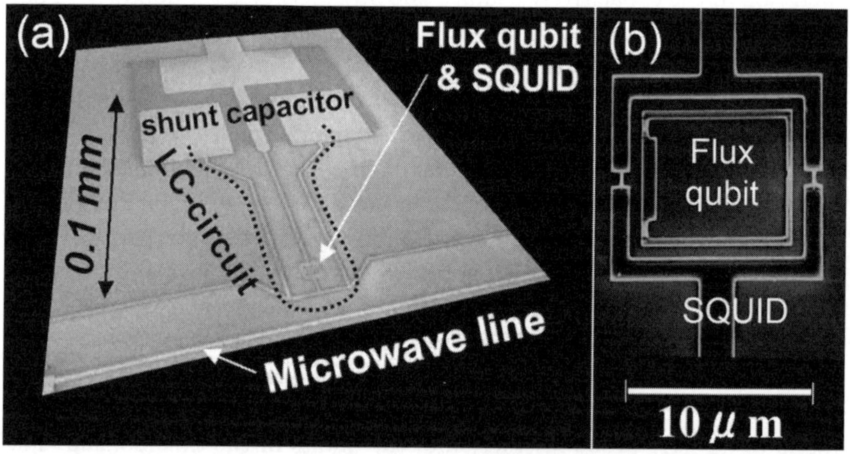

Fig. 12. Scannning Electron Micrograph of the sample measured at NTT-BRL Atsugi Labs. (a) SEM picture of a flux qubit together with near-by on-chip elements. Note that the size of the LC-circuit made of shunt capacitor and lead inductance is an order of 0.1 mm large. (b) Close-up view of the central part of the device. A flux qubit with a dc-SQUID; a quantum detector of the qubit state.

Very recently, the NTT group has successfully observed time domain vacuum Rabi oscillation in a macroscopic superconducting flux qubit and on-chip linear lumped element LC oscillator system[117]. SEM photograph of the sample is shown in Fig.12(a). The flux-qubit (Fig.12(b)) and the LC-oscillator is inductively coupled. Estimated coupling strength 0.2 GHz

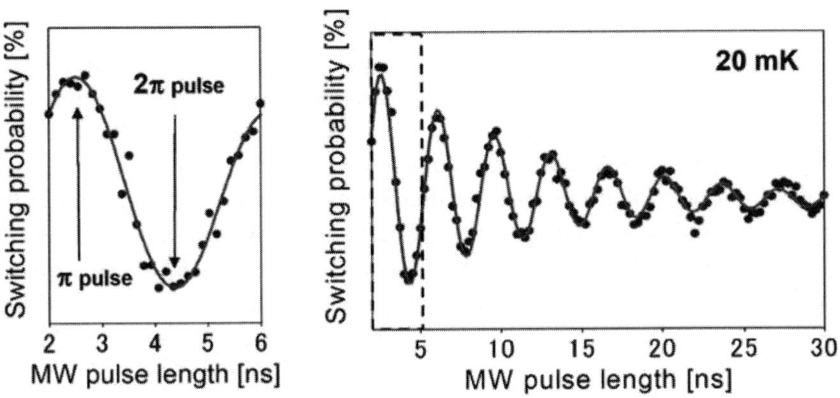

Fig. 13. Rabi oscillations of the flux-qubit during resonant (E = 14 GHz) microwave pulse, at temperature of 20 mK. The π and 2π-pulse for qubit control are determined from Rabi oscillations. The left figure is the close-up of the initial part of the Rabi oscillations framed by the broken line. The amplitude of the Rabi oscillation is ~25%.

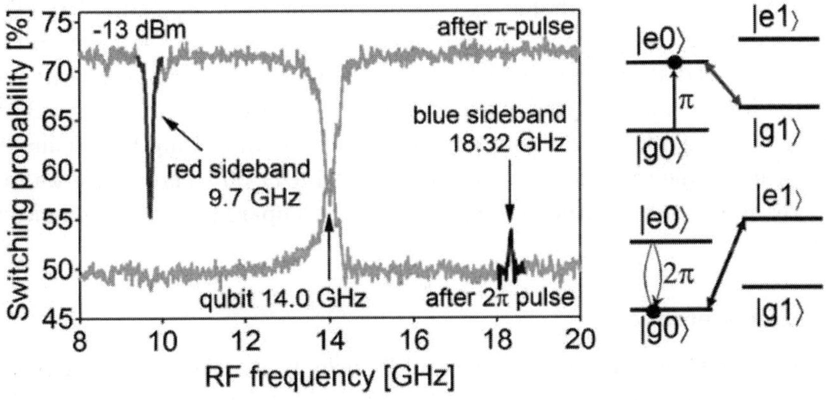

Fig. 14. Conditional spectroscopy observed in the qubit-LC oscillator coupled system. The qubit resonant frequency is fixed at E = 14 GHz. The lower (upper) trace is taken when the qubit is in the ground (excited) state. Only the blue- (red-) sideband is observed except for the qubit transition which can be understood by the level scheme of the coupled system.

is well in the strong coupling regeme. The resonance frequency of the LC-oscillator was observed as 4.3 GHz ($\sim$ 200 mK). Experiment was done with temperature of 20 mK, which was well below the smallest excitation energy including the LC-oscillator quantum. As shown in Fig.13, π, and

2π-pulse used in the experiment were determined from the Rabi oscillation. Figure 14 shows example of the conditional spectroscopy, when qubit energy was biased at 14 GHz. The red- and blue-sideband transition and also driven Rabi oscillations were clearly observed just like trapped ions interacting with their collective normal mode oscillation [118]. We have observed vacuum Rabi oscillations in time domain as follows. Preparing the $|e,0\rangle$ state by applying a π-pulse to the ground state $|g,0\rangle$ (we adopt the notation $|\text{qubit,Fock state of the LC-oscillator}\rangle$). Immediately after the π-pulse, applied dc-flux bias shift pulse brings the qubit to the resonant point where $|e,0\rangle$ and $|g,1\rangle$ have the same energy. Here, the rise/fall time of the dc-shift pulse should be adiabatic to the qubit i.e., longer than 0.1 ns, but non-adaibatic to the coupling between qubit and the LC-oscillator i.e., shorter than 5 ns. Then switching probability of the detecting SQUID, which is directly related to the excited state occupation of the qubit, as a function of the time period of resonance condition, we observe a decaying sinusoidal oscillation with frequency Ω_R between $|e,0\rangle$ and $|g,1\rangle$. By exciting the LC-oscillator very weakly during vacuum Rabi measurement pulse sequence, we confirmed gradual frequency crossover of the Rabi oscillation from Ω_R to expected higher frequency component $\sqrt{2}\Omega_R$ which is coming from the coherent oscillation between $|e,1\rangle$ and $|g,2\rangle$. This is the first direct evidence of the level quantization of macroscopic LC-resonator. By replacing a Rydberg atom to a flux-qubit and a high-Q cavity to an on-chip LC-oscillator, we will be able to perform cavity QED experiment on a chip where we can enjoy orders of magnitude stronger coupling compared to atomic quantum optics.

7. Decoherence

Coherence is the ability to interfere and is usually associated with the phase of a wave. In optics, light is called incoherent if its propagation can be described by geometrical optics alone [119]. The applicability of such a description clearly depends on the phenomenon being studied and on the scale of observation. The focusing of coherent laser light on large scales can, e.g., still be described by geometrical optics, the interference phenomena of the same laser beam in an interferometer require the use of a wave description. In quantum mechanics, the phase under consideration is usually the phase of the wave function and the incoherent limit of the theory is classical physics. Decoherence is the loss of coherence — in quantum mechanics it describes the transition from generically quantum to classical behavior. When describing decoherence, one thus has to clearly specify which quantum interference phenomenon (and which classical counterpart) is being studied. E.g. in a two-state-system (TSS) one can specify the "decoherence of free quantum oscillations" instead of just "decoherence". We will see that there are regimes when specific quantum phenomena, such as the formation of superpositions of classical states, are still observable, whereas others, such as real-time interference fringes, are already completely suppressed. In fact, this property of decoherence is important for the coexistence of classical and quantum description of matter e.g. in the large-scale classical description of molecules bound together by the quantum phenomenon of chemical bonding. The understanding of decoherence has huge impact on the general understanding of quantum mechanics [120–122].

Decoherence is an irreversible phenomenon including dissipation of energy and/or the generation of entropy. Consequently decoherence is not a generic part of elementary quantum mechanics based on the Schrödinger equation or its relativistic generalizations, which are all reversible at least in the sense of the CPT-theorem. This apparent contradiction is easily solved in large systems with a thermodynamic number of degrees of freedom: Such a system has a high number of levels and from a general initial state its wave function will follow a complex beating of oscillations whose frequencies are set by all possible transition frequencies $\omega_{nm} = (E_n - E_m)/\hbar$ between all combinations of levels n, m. It will return to its initial state after a time T which satisfies $T = 2p_{nm}\pi/\omega_{nm}$ for all n, m with a set of integers p_{nm}. This time is called Poincaré time and has very large values in thermodynamic systems with (quasi)continuous spectra. Thus, the reversibility cannot be observed in any reasonable experiment. On the other hand, there are always pairs of ω_{nm} and $\omega_{m'n'}$ very close together, such that coherent phenomena between states n and m are completely masked by those of states m', n' and can also not be observed on short time scales. Moreover, observing a system exactly with this precision in a well-defined way requires the repeated preparation of the same microscopic initial state, which is not possible for such a high number of degrees of freedom. In other words: Even though microscopic physics may be reversible and coherent, we are very often not able to observe it.

From the above discussion we can readily understand how coherence and reversibility vanish for large systems. Here, we are however mostly interested in small arrays of qubits, systems with very few (effective) degrees of freedom, and want to understand how they lose their coherence. The type of model invoked is closely related to the above example: The system is extend by coupling the quantum system to an environment (bath) containing a number of macroscopic degrees of freedom, whose detailed initial state is unknown except for its thermodynamical variables. Then one solves the dynamics of the full setup containing the quantum system and the bath. The solution will depend on the initial state of the bath and by tracing out the average over the ensemble of initial states of the bath, the effective reduced dynamics of the system under the influence of the bath is found. Baths of this kind always occur in nature: Generally, one has to be aware of the fact that the experimental machinery and the control and manipulation instruments can serve as a bath; in fact, even the electromagnetic vacuum is a bath to which energy can be emitted [123]. It is in particular starighforward to identify physical baths in solids: All the "unused" degrees of freedom, typically the lattice and electronic excitations, act as sources of decoherence for the qubit degrees of freedom. Recalling the preceding discussion of superconducting qubits, we note that the ubiquitous natural baths in a solid-state environment are suppressed: lattice vibrations are frozen out and the electronic excitations are gapped: Thus, superconductors are good candidates for maintaining coherence in solid-state qubits.

Due to the necessity to introduce an environment, quantum systems suffering from decoherence are often called "open quantum systems".

The transition to classical physics manifests itself in at least three ways, which are related but not equivalent. These cases will now be illustrated invoking a system described by a two-state Hamiltonian of the form eq. (11).

Firstly, a system can be by all means coherent and sufficiently isolated from the environment, but the coherent phenomenon manifests itself on an unobservable scale. E.g., we would not be able to observe coherence fringes in the propagation of a ping-pong ball, because the de Broglie wave length of the ball at any reasonable velocity is too small to be observed (at $v = 1\text{m/s}$ it is on the order of 10^{-30}m). In eq. (11), this would occur if the off-diagonal matrix element is so small that one cannot keep $\epsilon = 0$ with sufficient precision and hence cannot prepare superpositions, or if $\epsilon \gg \Delta$ such that coherent oscillations are too fast to be observed and/or too small in amplitude. This can be attributed to the fact that the quantum fluctuations leading to the coherent coupling of the classical states (here, the eigenstates of $\hat{\sigma}_z$) are too small. For tunneling systems such as the flux qubit, Δ is proportional to the overlap of the basis states in some double-well potential [74]. We can thus associate a small Δ with the fact that the basis states are too strongly separated for coherent tunneling. They are hence too distinct - very similar to the fact that the wave functions "alive" and "dead" of Schrödinger's cat [124] do not overlap. This type of vanishing quantum effects is usually not referred to as decoherence but as macroscopic distinction [87] and has also been termed "false decoherence" [86].

Secondly, if the system has an appropriate effective Hamiltonian with a sufficiently large off-diagonal matrix element it would be able to show all kinds of quantum coherent interference phenomena leading to fringes or temporal oscillations, such as the coherent oscillations of a two-state system which is initially *not* prepared in an eigenstate of the Hamiltonian eq. (11). Due to the influence of the environment these oscillations lose their phase and are suppressed. This can be understood in an ensemble-average: The system plus the bath propagate together and display a complex beating pattern. As the bath is prepared in somewhat different states in each member of the ensemble, this propagation looks differently for each realization and the quantum subsystem, from a unique initial state, accumulates more and more differences between one realization and the other, until ensemble-averaged quantum properties die out. This ensemble can well be taken in time, by repeating experiments with identical initial states of the quantum subsystem. This phenomenon is called "dephasing". It does not necessarily involve the exchange of energy with the environment. On the other hand, at long times we expect the system to go into a thermal state described by a diagonal Boltzmann-type density matrix. This process involves energy exchange with the environment and is called relaxation. It takes place if there are no special selection rules and *does* necessarily also lead to dephasing. The thermal state can still involve superpositions of the basis states, if the Hamiltonian has off-diagonal terms. All in all this type of dynamics is called ("generic") decoherence.

Thirdly even if the system Hamiltonian eq. (11) has substantial off-diagonal elements, coherent tunneling between the classical states can be blocked on all time scales by interaction with the environment. In fact, the system gets dressed by fast degrees of freedom in the environment, i.e. the system states $|\psi_n\rangle$ have to be replaced by combined states $|\psi_n\rangle_{\rm eff} = |\psi_n\rangle \otimes |ENV_n\rangle$, where $|ENV_n\rangle$ are the lowest-energy states of the bath under the condition that the quantum system is in state $|\psi_n\rangle$. This type of dressing leads to a renormalization of the system Hamiltonian. E.g., in eq. (11), Δ gets renormalized to $\Delta_{\rm eff}$ when one couples a bath of oscillators to $\hat{\sigma}_z$ [92,125]. Such renormalization effects are known throughout physics as Lamb shift, Franck-Condon effect etc. , all of which describe the introduction of an effective Hamiltonian whose matrix elements are different from the original ones due to the interaction with some environment.

If the bath is infinite and has sufficient spectral weight at all frequencies, the system can undergo a dissipative phase transition [92] which leads to $\Delta_{\text{eff}} = 0$, making the system completely classical and localized, similar to Anderson's orthogonality catastrophe [126] in Fermionic systems. In fact, the dynamics described in the previous paragraph is always governed by H_{eff}. In particular, there are no superpositions of classical states left in a thermal mixture. Another way to interpret this is, that the system builds up entanglement with the environment, i.e. any superposition of different $|\psi_n\rangle_{\text{eff}}, |\psi_m\rangle_{\text{eff}}$ is entangled. If the entanglement is complete, such that the $|ENV_n\rangle$ are mutually orthogonal, the system cannot tunnel any more between the classical states $_{\text{eff}}\langle\psi_n|\psi_{m\neq n}\rangle_{\text{eff}} \to 0$. This is analogous to the first scenario: There the system states themselves are macroscopically distinct, here only the entangled system plus environment states are.

7.1. *Methods*

The discussion in the previous section already indicates the importance of studying noise for understanding decoherence [127,128]. In fact, parts of our results on decoherence can be understood from noise theory alone without a detour via an environmental Hamiltonian. This can be illustrated by the two-state system described by (11), where ϵ is noisy, $\epsilon = \epsilon_0 + \delta\epsilon(t)$. We can readily solve the associated Schrödinger equation. Even more intuitively we can describe the state of the system by the expectation values of the three components of the spin, and the Schrödinger equation becomes the classical equation of motion of magnetic moment in a fluctuating magnetic field. For each realization of the noise, the system behaves coherently, but the coherent evolution has a noisy component. Averaging over the noise then leads to incoherent evolution. Specifically, two key results can be derived on that level, assuming that $\delta\epsilon$ can be treated perturbatively [129,130]: The relaxation rate is proportional to the noise spectral density $S = \langle\{\delta\epsilon(t), \delta\epsilon(0)\}\rangle\, \omega/2$ at a frequency corresponding to the level splitting $E = \sqrt{\epsilon^2 + \Delta^2}$

$$\Gamma_r = \frac{1}{T_1} = \frac{\Delta^2}{E^2} S(E). \tag{12}$$

Flip-less decoherence contains a zero energy exchange (zero frequency) contribution as well [131,132,30]

$$\Gamma\phi = \frac{1}{T_2} = \Gamma_r/2 + \frac{\epsilon^2}{E^2} S(0). \tag{13}$$

These results can be intuitively interpreted: The system can relax by dissipating all its energy E into an environmental Boson. Due to the weakness of the coupling, multi-Boson processes are strongly supressed. The relaxation also dephases the state. Moreover, dephasing can occur due to the coupling to low-frequency modes which do not change the energy of the system. These expressions for relaxation and dephasing have also been found by studying the Hamiltonian of our qubit coupled to a damped oscillator, using a Markovian master equation approach by Tian *et al.* [133] (based on work by Garg *et al.* [134]).

The expressions (12) and (13) have prefactors $\left(\frac{\Delta}{E}\right)^2$ and $\left(\frac{\epsilon}{E}\right)^2$ that depend on the tunnel splitting Δ and the energy bias ϵ. These factors correspond to the angles between noise and eigen states usually introduced in NMR [135] and account for the effect that the qubit's magnetic dipole radiation is strongest where the flux in the qubit $\Phi_q = \frac{1}{2}\Phi_0$ (i. e. $\left(\frac{\Delta}{E}\right)$ maximal), and that the level separation ν is insensitive to flux noise at this point (i. e. $\frac{\partial \nu}{\partial \epsilon} = \left(\frac{\epsilon}{\nu}\right) \approx 0$). One should know and control $S(\omega)$ at the frequency $E/\hbar$ for controlling the relaxation, and at low frequencies for controlling the dephasing. In what follows we will calculate the noise properties of a few typical experimental environments, and calculate how the noise couples to the qubit. In fact, these results are at the core of the choice of an optimum working points described in section 5.3: If $S(0)$ from a specific environment is particularily destructive, e.g. in $1/f$ charge noise, we make sure that that environment couples to the qubit in a way that is perpendicular on the Bloch sphere such that the flipless term vanishes. A lucid review of this quantum noise approach can be found in Ref. [127].

Such methods for describing decoherence by averaging over semiclassical noise will generally fail whenever entanglement between system and bath becomes significant. This happens at large time scales, but also at stronger coupling beyond perturbation theory. This failure is due to the effect that under these conditions the system also has a pronounced influence on the bath, which changes its dynamics and from there acts back on the system. This is not captured if the noise is calculated for an a priori given (thermal) bath state.

The more conventional methods for describing decoherence rely on studying the reduced density matrix of the quantum system. This density matrix is obtained from the full density matrix of system and environment by tracing out the environment as described above. The reduced density matrix describes an ensemble of systems, i.e. the results obtained from it can be compared to results averaged over huge collections of quantum systems or a repeated experiment with identical initial conditions of the quantum system at each attempt (time-ensemble). The behavior of a *single* realization is *not* predicted, similar to the statistical description of quantum measurements [136], see above. In most cases the dynamics of the reduced density matrix can be described by a generalized master-equation with memory, i.e. by a linear integro-differential equation containing all the evolution in the past, which falls back onto the Liouville equation if the coupling to the bath is taken away [137–140]. Even the path-integral descriptions for open quantum systems developed from the 80s on have recently been cast in a master-equation form [140]. Master equations which are local in time are called *Markovian* [141–145]. Many approaches have been formulated within the Born approximation, which only contains the coupling to the bath in lowest, quadratic, order [146]. The simplest example for such a master equation is the phenomenological Bloch equation from NMR [147,131], which introduces longitudinal relaxation rates (relaxation of the spin component parallel to the magnetic field, corresponding to energy relaxation) and transversal rates corresponding to dephasing. These equations have first been phenomenologically introduced in the picture of classical spins without having decoherence in mind. They have been further developed in NMR and can be microscopically [148,131] derived under suitable Markov assumptions.

There are other methods to describe open quantum systems, many of which are essentially numerical: The quantum jump or trajectory method [149,150] and the Bayesian formalism [151] rely on additional assumptions and have the advantage that they stay within a wave function representation, which is advantageous for huge systems. Various renormalization group approaches [152–154] on the other hand are very precise and include the regime of large coupling; however, up to now they have difficulties in treating nonequilibrium or driven situations.

7.2. *Heat bath environments*

So far we have given a number of examples for baths which are physically quite different, but have not detailed how to model them. It turns out that many baths can be modelled using a few general models: In the thermodynamic limit at equilibrium whenever the central limit theorem holds, systems are *Gaussian*, i.e. the distribution of values for collective variables X is of Gaussian form and can be fully described by the two-time correlation function $\langle X(t)X(0)\rangle$. All higher cumulants are zero. Such Gaussian models can be universally described as a bath of harmonic oscillators, and collective variables can be written as a linear combination of the oscillator coordinates [67,92,140]. In order to determine the appropriate couplings and distribution of the oscillators one introduces a spectral density $J(\omega)$ of the oscillators which is determined from the specific properties of the model under consideration. This matching to the physical model is usually done by comparing either the classical friction induced by the environment, or the noise correlation function, and will be detailed later on. Then the model of system plus bath with the appropriate $J(\omega)$ can be solved quantum-mechanically as described above.

Oscillator bath models work well e.g. for phonon and photon baths such as electromagnetic noise. In the latter case, the (hard-core) bosons of the environment are electron-hole pairs. In many cases, the hard-core correction does not play any role and is of measure zero in the thermodynamic limit. For superconducting qubits of many kinds, the environmental spectral function $J(\omega)$ can be readily determined from either the *noise* or the *classical friction* induced by the environment. We would like to illustrate this procedure now for the case of flux qubits.

Any linear electromagnetic environment can be described by an effective impedance Z_{eff}. If the circuit contains Josephson junctions below their critical current, they can be included through their kinetic inductance $L_{\text{kin}} = \Phi_0/(2\pi I_c \cos\bar{\phi})$ (see section 3), where $\bar{\phi}$ is the average phase drop across the junction. The circuitry disturbs the qubit through its Johnson-Nyquist noise, which has Gaussian statistics and can thus be described by an effective Spin-Boson model [92]. In this model, the properties of the oscillator bath which forms the environment are characterized through a spectral function $J(\omega)$, which can be derived from the external impedance. To stay within an all-electrical analogy: We replace the resistors in the setup by infinite LC lines and perform a normal mode transformation to decouple all oscillators. Note, that other nonlinear elements such as tunnel junctions which can produce non-Gaussian shot noise are generically *not* covered by oscillator bath models.

As explained above, the flux noise from an external circuit leads to $\epsilon = \epsilon_0 + \delta\epsilon(t)$. For being able to treat system-bath entanglement, we write out a phenomenolgical Hamiltonian, the spin boson model,

$$H_{SB} = \frac{\epsilon}{2}\hat{\sigma}_z + \frac{\Delta}{2}\hat{\sigma}_x + \frac{1}{2}\hat{\sigma}_z \sum_i \lambda_i (a_i + a_i^\dagger) + \sum_i a_i^\dagger a_i \tag{14}$$

We again parametrize the noise $\delta\epsilon(t)$ by its semiclassical power spectrum

$$S(\omega) = \langle\{\delta\epsilon(t), \delta\epsilon(0)\}\rangle\, \omega = \hbar^2 J(\omega) \coth(\hbar\omega/2k_B T). \tag{15}$$

Here, $J(\omega)$ is formally defined as $J(\omega) = \sum_i \lambda_i^2 \delta(\omega - \omega_i)$. We will explain how $J(\omega)$ can be obtained for electromagnetic environments in section 7.3. Firstly, we would like to outline an alternative approach pioneered by Leggett [155], where $J(\omega)$ is derived from the classical friction induced by the environment. In reality, the combined system of SQUID and qubit will experience fluctuations arising from additional circuit elements at different temperatures, which can be treated in a rather straightforward manner.

7.3. *Decoherence due to the Electromagnetic Environment*

7.3.1. *Characterizing the Environment from Classical Friction*

We study a DC-SQUID in an electrical circuit as shown in Fig. 3. It contains two Josephson junctions with phase drops denoted by $\gamma_{1/2}$. We start by looking at the average phase $\gamma_{\rm ex} = (\gamma_1 + \gamma_2)/2$ across the read-out SQUID. Starting from the generalization of the RCSJ model to arbitrary shunt admittances Y, we find the equation of motion

$$2C_J \frac{\Phi_0}{2\pi} \ddot{\gamma}_{\rm ex} = -2I_{\rm c,0} \cos(\gamma_{\rm i}) \sin \gamma_{\rm ex} + I_{\rm bias} - \frac{\Phi_0}{2\pi} \int dt' \dot{\gamma}_{\rm ex}(t') Y(t-t'). \quad (16)$$

Here, $\gamma_{\rm in} = (\gamma_1 - \gamma_2)/2$ is the dynamical variable describing the circulating current in the loop which is controlled by the flux, $I_{\rm bias}$ is the bias current imposed by the source, $Y(\omega) = Z^{-1}(\omega)$ is the admittance in parallel to the whole SQUID and $Y(\tau)$ its Fourier transform. The SQUID is described by the junction critical currents $I_{\rm c,0}$ which are assumed to be equal, and their capacitances $C_{\rm J}$. We now proceed by finding a static solution which sets the operation point $\gamma_{\rm in/ex,0}$ and small fluctuations around them, $\delta\gamma_{\rm in/ex}$. The static solution reads $I_{\rm bias} = I_{\rm c,eff} \sin \gamma_{\rm ex,0}$ where $I_{\rm c,eff} = 2I_{\rm c,0} \cos \gamma_{\rm in,0}$ is the effective critical current of the SQUID. Linearizing Eq. (16) around this solution and Fourier-transforming, we find that

$$\delta\gamma_{\rm ex}(\omega) = \frac{2\pi I_{\rm b} \tan \gamma_{\rm in,0} Z_{\rm eff}(\omega)}{i\omega\Phi_0} \delta\gamma_i(\omega) \quad (17)$$

where $Z_{\rm eff}(\omega) = \left(Z(\omega)^{-1} + 2i\omega C_{\rm J} + (i\omega L_{\rm kin})^{-1}\right)^{-1}$ is the effective impedance of the parallel circuit consisting of the $Z(\omega)$, the kinetic inductance of the SQUID and the capacitance of its junctions. Neglecting self-inductance of the SQUID and the (high-frequency) internal plasma mode, we can straightforwardly substitute $\gamma_{\rm in} = \pi\Phi/\Phi_0$ and split it into $\gamma_{\rm in,0} = \pi\Phi_{\rm x,S}/\Phi_0$ set by the externally applied flux $\Phi_{\rm x,S}$ through the SQUID loop and $\delta\gamma_{\rm i} = \pi M_{\rm SQ} I_{\rm Q}/\Phi_0$ where $M_{\rm SQ}$ is the mutual inductance between qubit and the SQUID and $I_{\rm Q}(\vec{\varphi})$ is the circulating current in the qubit as a function of the junction phases, which assumes values $\pm I_{\rm p}$ in the classically stable states.

In order to analyze the backaction noise of the SQUID onto the qubit in the two-state approximation, we have to get back to its full, continuous description we started out from in section 5.2, the classical dynamcis. These are equivalent to a particle, whose coordinates are the two independent junction phases in the three-junction loop, in a two-dimensional potential

$$\vec{C}(\Phi_0/2\pi)^2\ddot{\vec{\varphi}} = -\nabla U(\vec{\varphi}, \Phi_{x,q} + I_S M_{SQ}). \tag{18}$$

The details of this equation are explained in Ref. [73] and in section 5.2. $\vec{C}$ is the capacitance matrix describing the charging of the Josephson junctions in the loop, $U(\vec{\varphi})$ contains the Josephson energies of the junctions as a function of the junction phases and I_S is the ciculating current in the SQUID loop. The applied flux through the qubit Φ_q is split into the flux from the external coil $\Phi_{x,q}$ and the contribution form the SQUID. Using the above relations we find

$$I_S M_{SQ} = \delta\Phi_{cl} - 2\pi^2 M_{SQ}^2 I_B^2 \tan^2 \gamma_{in,0} \frac{Z_{eff}}{i\omega\Phi_0^2} I_Q \tag{19}$$

where $\delta\Phi_{cl} \simeq M_{SQ} I_{c,0} \cos\gamma_{ex,0} \sin\gamma_{in,0}$ is the non-fluctuating back-action from the SQUID.

From the two-dimensional problem, we can now restrict ourselves to the one-dimensional subspace defined by the preferred tunneling direction shown in Fig. 4 [73], which is described by an effective phase φ. The potential restricted on this direction, $U_{1D}(\varphi)$ has the form of a double well [92,140] with stable minima situated at $\pm\varphi_0$. In this way, we can expand $U_{1D}(\varphi, \Phi_q) \simeq U(\varphi, \Phi_q, x) + I_Q(\varphi) I_Q M_{SQ}$. Approximating the phase-dependence of the circulating current as $I_Q(\varphi) \approx I_p \varphi/\varphi_0$ where I_p is the circulating current in one of the stable minima of φ, we end up with the classical equation of motion of the qubit including the backaction and the friction induced from the SQUID

$$\left[-C_{eff}\left(\frac{\Phi_0}{2\pi}\right)^2 \omega^2 + 2\pi^2 M_{SQ}^2 I_{bias}^2 \tan^2 \gamma_{in,0} \frac{Z_{eff} I_p^2}{i\varphi_0 \omega \Phi_0^2}\right]\varphi$$
$$= -\partial\varphi U_{1D}(\varphi, \Phi_{x,q} + \delta\Phi_{cl}). \tag{20}$$

We encode this form using the Fourier transform of the differential operators as $D(\omega)\varphi(\omega) = -\partial U/\partial\varphi$. We can use the prescription given in [155] and identify the spectral function for the continuous, classical model as $J_{cont} = \mathrm{Im} D(\omega)$. From there, we can do the two-state approximation for the particle in a double well [140] and find $J(\omega) = J_{cont}$ in analogy to [156]

$$J(\omega) = \frac{(2\pi)^2}{\hbar\omega}\left(\frac{M_{SQ} I_p}{\Phi_0}\right)^2 I_{bias}^2 \tan^2\left(\frac{\pi\Phi}{\Phi_0}\right) \mathrm{Re}\{Z_{eff}(\omega)\}. \tag{21}$$

This equation has a remarkable consequence: It predicts a well-defined "off"-state at $I_{\text{bias}} = 0$, i.e. as for any good detector, the backaction can be switched off in order to prevent the Zeno effect. In practice, one has to take into account the double-layer structure, so either the preferred working point is at finite current, or the double layer is shorted by a fourth junction[90], see also section 5.3.

Note that, even though we have assumed a shunt only across the whole device, this method can be extended to describe shunts at the individual SQUID junctions as well [157].

7.4. *Qubit Dynamics under the Influence of Decoherence*

From $J(\omega)$, we can analyze the dynamics of the system by studying the reduced density matrix, i.e. the density matrix of the full system where the details of the environment have been integrated out, by a number of different methods. The low damping limit, $J(\omega)/\omega \ll 1$ for all frequencies, is most desirable for quantum computation. Thus, the energy-eigenstates of the qubit Hamiltonian are the appropriate starting point of our discussion. In this case, the relaxation rate Γ_r (and relaxation time τ_r) are determined by the environmental spectral function $J(\omega)$ at the frequency of the level separation E of the qubit by equations (12) and (13), These expressions have been derived in the context of NMR [135] and recently been confirmed by a full path-integral analysis [132].

For performing efficient measurement, one can afford to go to the strong damping regime. A well-known approach to this problem, the noninteracting blip approximation (NIBA) has been derived in Ref. [92]. This approximation gives good predictions at degeneracy, $\epsilon = 0$. At low $|\epsilon| > 0$ it contains an artifact predicting incoherent dynamics even at weak damping. At high bias, $\epsilon \gg \Delta$ and at strong damping, it becomes asymptotically correct again. We will not detail this approach here more, as it has been extensively covered in [92,140].

If $J(\omega)$ is not smooth but contains strong peaks the situation becomes more involved: At some frequencies, $J(\omega)$ may fall in the weak and at others in the strong damping limit. In some cases, whern $J(\omega) \ll \omega$ holds at least for $\omega \leq \Omega$ with some $\Omega \gg \nu/\hbar$, this can be treated approximately: one can first renormalize Δ_{eff} through the high-frequency contributions [92] and then perform a weak-damping approximation from the fixed-point Hamiltonian. This is detailed in Ref. [158]. In the general case, more involved methods such as flow equation renormalization [159,160], QUAPI [161] or the complex environments [162] approach have to be used.

7.5. *Application: Engineering the Measurement Apparatus*

From Eq. (21) we see that engineering the decoherence induced by the measurement apparatus essentially means engineering Z_{eff}. We specifically focus on the contributions due to the measurement apparatus. In this section, we are going to outline and compare several options suggested in literature. We assume a perfect current source that ramps the bias current I_{bias} through the SQUID. The fact that the current source is non-ideal, and that the wiring to the SQUID chip has an impedance is all modeled by the impedance $Z(\omega)$. The wiring can be engineered such that for a very wide frequency range the impedance $Z(\omega)$ is on the order of the vacuum impedance, and can be modeled by its real part R_l. It typically has a value of 100 Ω.

It has been suggested [30] to overdamp the SQUID by making the shunt circuit a simple resistor $Z(\omega) = R_{\text{S}}$ with $R_{\text{S}} \ll \sqrt{L_{\text{kin}}/2C_{\text{J}}}$. This is inspired by an analogous setup for charge qubits, [163]. Following the parameters given in [156], we find $\alpha R = 0.08\Omega$ and $\omega_{\text{LR}}/R = 8.3\text{GHz}/\Omega$.

Next, we consider a large superconducting capacitive shunt (Fig. 15a, as implemented in Refs. [75,164]). The C shunt only makes the effective mass of the SQUID's external phase γ_{ex} very heavy. The total impedance $Z_{\text{eff}}(\omega)$ and $J(\omega)$ are modeled as before, see Fig. 16. As limiting values, we find

$$\text{Re}\{Z_{\text{eff}}(\omega)\} \approx \begin{cases} \frac{\omega^2 L_J^2}{R_l}, & \text{for } \omega \ll \omega_{LC} \\ R_l, & \text{for } \omega = \omega_{LC} \\ \frac{1}{\omega^2 C_{sh}^2 R_l}, & \text{for } \omega \gg \omega_{LC} \end{cases} \tag{22}$$

We can observe that this circuit is a weakly damped LC-oscillator and it is clear from (12) and (21) that one should keep its resonance frequency $\omega_{\text{LC}} = 1/\sqrt{L_{\text{J}}C_{\text{sh}}}$, where $\text{Re}\{Z_{\text{eff}}(\omega)\}$ has a maximum, away from the qubit's resonance $\omega_{\text{res}} = \nu/\hbar$. This is usually done by chosing $\omega_{\text{LC}} \ll \omega_{\text{res}}$. For a C-shunted circuit with $\omega_{LC} \ll \omega_{res}$, this yields for $J(\omega \approx \omega_{\text{LC}})$

$$J(\omega) \approx \frac{(2\pi)^2}{\hbar\omega^3}\left(\frac{MI_{\text{p}}}{\Phi_0}\right)^2 I_{\text{bias}}^2 \tan^2\left(\frac{\pi\Phi}{\Phi_0}\right)\frac{1}{C_{sh}^2 R_l} \tag{23}$$

One has to be aware of the fact that at these high switching current values the linearization of the junction as a kinetic inductor may underestimate the actual noise. In that regime, phase diffusion between different minima of the washboard potential also becomes relevant and changes the noise properties [165,166]. As a complementary perspective, the C-shunted SQUID can be seen as a, generally nonlinear, cavity with small leakage rate [114,162].

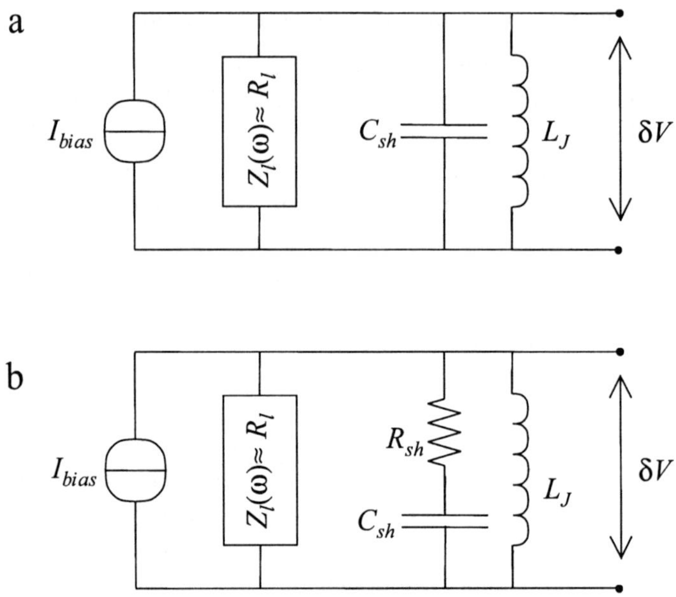

Fig. 15. Circuit models for the C-shunted DC-SQUID (a) and the RC-shunted DC-SQUID (b). The SQUID is modeled as an inductance L_J. A shunt circuit, the superconducting capacitor C_{sh} or the R_{sh}-C_{sh} series, is fabricated on chip very close to the SQUID. The noise that couples to the qubit results from Johnson-Nyquist voltage noise δV from the circuit's total impedance Z_{eff}. Z_{eff} is formed by a parallel combination of the impedances of the leads Z_l, the shunt and the SQUID, such that $Z_{\mathrm{eff}}^{-1} = 1/Z_l + 1/(R_{sh} + 1/i\omega C_{sh}) + 1/i\omega L_J$, with $R_{sh} = 0$ for the circuit (a)

As an alternative we will consider a shunt that is a series combination of a capacitor and a resistor (Fig. 15b) (RC-shunted SQUID). The RC shunt also adds damping at the plasma frequency of the SQUID, which is needed for realizing a high resolution of the SQUID readout (i. e. for narrow switching-current histograms) [165]. The total impedance $Z_t(\omega)$ of the two measurement circuits are modeled as in Fig. 15. For the circuit with the RC shunt

$$\mathrm{Re}\{Z_t(\omega)\} \approx \begin{cases} \frac{\omega^2 L_J^2}{R_l}, & \text{for } \omega \ll \omega_{LC} \\ \leq R_l, & \text{for } \omega = \omega_{LC} \ll \frac{1}{R_{sh}C_{sh}} \\ R_l//R_{sh}, & \text{for } \omega = \omega_{LC} \gg \frac{1}{R_{sh}C_{sh}} \\ R_l//R_{sh}, & \text{for } \omega \gg \omega_{LC} \end{cases} \tag{24}$$

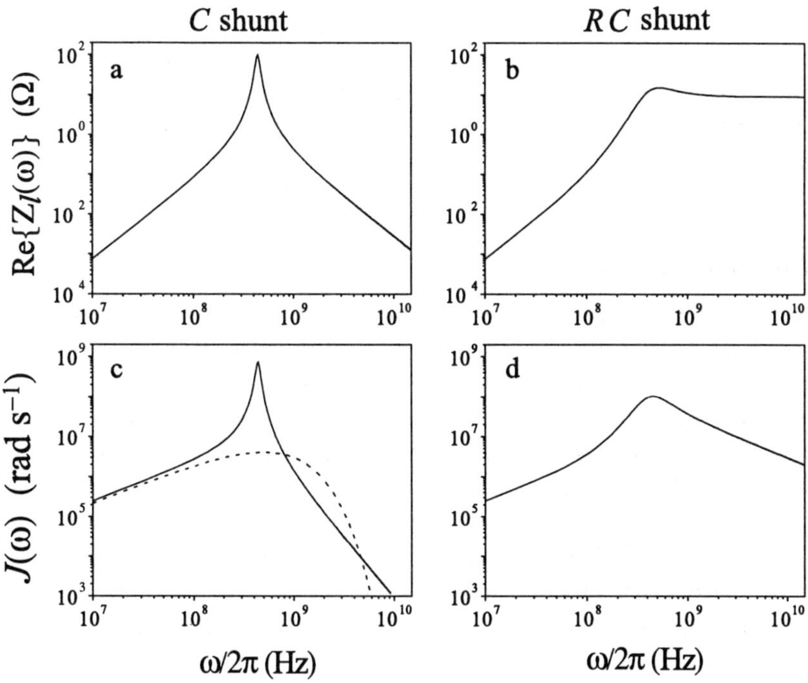

Fig. 16. A typical $\text{Re}\{Z_t(\omega)\}$ for the C-shunted SQUID (a) and the RC-shunted SQUID (b), and corresponding $J(\omega)$ in (c) and (d) respectively. For comparison, the dashed line in (c) shows a simple Ohmic spectrum, $J(\omega) = \alpha\omega$ with exponential cut off $\omega_c/2\pi = 0.5$ GHz and $\alpha = 0.00062$. The parameters used here are $I_{\text{p}} = 500$ nA and $T = 30$ mK. The SQUID with $2I_{co} = 200$ nA is operated at $f = 0.75\,\pi$ and current biased at 120 nA, a typical value for switching of the C-shunted circuit (the RC-shunted circuit switches at higher current values). The mutual inductance $M = 8$ pH (i. e. $MI_{\text{p}}/\Phi_0 = 0.002$). The shunt is $C_{\text{sh}} = 30$ pF and for the RC shunt $R_{\text{sh}} = 10$ Ω. The leads are modeled by $R_l = 100\ \Omega$

The difference mainly concerns frequencies $\omega > \omega_{LC}$, where the C-shunted circuit has a stronger cutoff in $\text{Re}\{Z_{\text{eff}}(\omega)\}$, and thereby a relaxation rate, that is several orders lower than for the RC-shunted circuit. Given the values of $J(\omega)$ from Fig. 16, one can directly see that an RC-shunted circuit with otherwise similar parameters yields at $\omega_{res}/2\pi = 10$ GHz relaxation times that are about four orders of magnitude shorter.

Let us remark that a different parameter regime of the RC shunt has been studied in Ref. [157]. Here, the RC shunts are higher in resistance and allow to enter the quantum tunneling regime for the SQUID. Also, they are not applied across the device, but to the individual junctions. The steering of the minimum width and the histograms has been studied in great detail and good agreement between theory and experiment has been obtained. However, it turned out that the best device performance will be achieved for $R = 0$, a pure C-shunt.

7.6. *Discrete noise*

Not all environments can be described this way. A prominent counterexample are localized modes with a bounded spectrum such as spins [167] and structural fluctuations generating classical telegraph noise [168–171], which are inherently non-Gaussian. One expects from the central limit theorem, that large ensembles of such fluctuators usually behave Gaussian again [172,130], but details of this transition are only partially understood at present [169,173].

Most research has concentrated on decoupling devices from external noise sources such as electromagnetic noise generated by control and measurement apparatus [156]. On the other hand, there inevitably are internal noise sources because the fabrication of gates, tunnel junctions, and other functional components creates defects in the underlying crystal. Prominent examples of such defects are background charges in charge-based devices or cricital current fluctuations in flux-based devices [174,175]. A clear signature of such defects is telegraph noise in the case of a few defects or $1/f$-noise in the case of a larger ensemble [176]. With the growing success in engineering the electromagnetic environment, these defects are becoming more and more the key limiting sources of decoherence.

Such defects do not fall in the large class of noise sources that can be approximated well as a bosonic bath, and this fact complicates their analysis. Localized noise sources with bounded spectra like the defects in which we are interested produce noise that is significantly non-Gaussian. Theories treating large ensembles of non-Gaussian noise sources have been presented [168,167]. However, with the ongoing improvement in nanofabrication technology, it is realistic to consider the case where non-Gaussian noise sources are reduced down to only a single one or a few per device. This is the case we treat here, and thus the defects find a more realistic representation as a small set of bistable fluctuators (henceforth abbreviated bfls). In principle, this approach can be extended to larger sets of bfls with a range of different mean switching times (*e.g.*, an ensemble with an exponential distribution of switching times that produces $1/f$-noise [99,173,171]).

7.7. *Model of the bistable fluctuator in its semiclassical limit*

We describe the bfl-noise influenced evolution of the qubit in its semiclassical limit by using a stochastic Schrödinger equation [177,178] with the time-dependent effective Hamiltonian

$$H_q^{\rm eff}(t) = H_{\rm q} + H_{\rm noise}(t) \tag{25}$$

$$H_{\rm q} = \hbar\epsilon_{\rm q}\hat{\sigma}_{\rm z}^{\rm q} + \hbar\Delta_{\rm q}\hat{\sigma}_{\rm x}^{\rm q} \tag{26}$$

$$H_{\rm noise}(t) = \hbar\alpha\ \hat{\sigma}_{\rm z}^{\rm q}\ \xi_{\rm bfl}(t) \tag{27}$$

where $\epsilon_{\rm q}$ and $\Delta_{\rm q}$ define the free (noiseless) qubit dynamics. $\xi_{\rm bfl}(t)$ denotes a function randomly switching between ± 1 (see Fig. 17), which represents a telegraph noise signal. The switching events follow a symmetrical Poisson process, *i.e.*, the probabilities of the bfl switching from $+1$ to -1 or -1 to $+1$ are the same and equal in time. The Poisson process is characterized by the mean time separation $\tau_{\rm bfl}$ between two bfl flips. The coupling amplitude to the qubit in frequency units is α.

Starting with an arbitrary initial state of the qubit, represented by some given point on the Bloch sphere, we can numerically integrate the corresponding stochastic differential equation and obtain the corresponding random walk on the Bloch sphere

$$\vec{\sigma}(t) = {\rm T}\exp\left(-i/\hbar\int_0^t H_{\rm q}^{\rm eff}(s)\,ds\right)\vec{\sigma}(0) \tag{28}$$

with T denoting the usual time-ordering operator.

7.8. *Bang-bang control protocol*

We propose to reduce the influence of the bfl-noise by applying to the qubit a continuous train of π-pulses along the $\sigma_{\rm x}$-axis. This refocusing pulse scheme essentially corresponds to the standard quantum bang-bang procedure [179–181] or the Carr-Purcell-Gill-Meiboom echo technique from NMR [182]. For technical convenience, we consider the π-pulses to be of infinitesimal duration. This simplification is not crucial as described in[183]. The pulses are assumed to be separated by a constant time interval $\tau_{\rm bb}$. The mean separation $\tau_{\rm bfl}$ between two bfl flips is assumed to be much longer than $\tau_{\rm bb}$. For theoretical convenience, we also assume that $\tau_{\rm bfl}$ is shorter than the free precession period of the qubit. This too is not a crucial restriction. (It can always be overcome by changing to a co-precessing frame.)

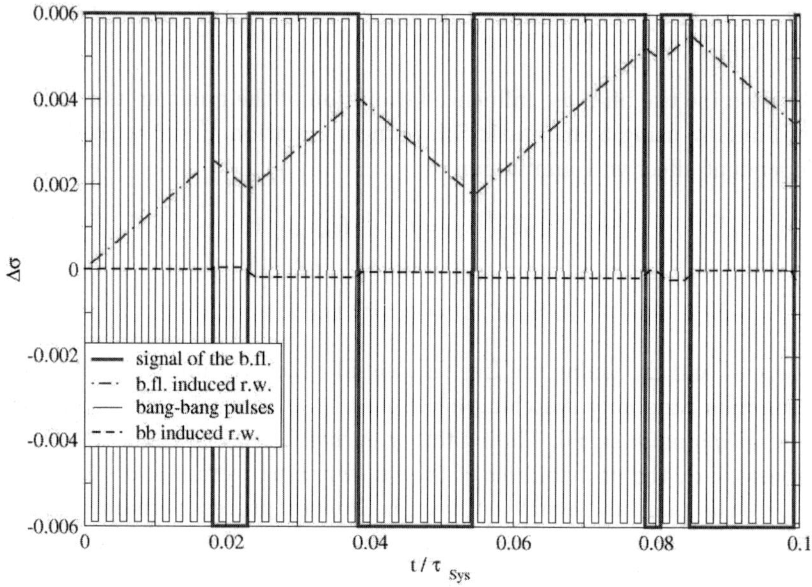

Fig. 17. Schematic plot of a typical Poisonian bfl noise signal and its resulting random walk behavior (in the limit of small deviations). The periodic fast switching step function represents a bang-bang pulse with a time scale ratio: $\tau_{\text{bfl}}/\tau_{\text{bb}} = 10$ and yields a quite smaller random walk step-length. $\tau_{\text{Sys}} = \frac{\pi}{\sqrt{\epsilon_{\text{q}}^2+\Delta_{\text{q}}^2}}$ denotes the evolution period of the qubit in the noiseless case.

Qualitatively, bang-bang control works as follows. Since $\tau_{\text{bb}} \ll \tau_{\text{bfl}}$, it is usually the case that the bfl does not flip during the time between two bang-bang pulses that flip the qubit. In this way, the bang-bang pulses average out the influence of $H_{\text{noise}}(t)$. In fact, the refocusing scheme fully suppresses the σ_z-term of the static Hamiltonian (26 (compare Fig. 21); but this turns out to be no crucial obstacle to universal quantum computation. As one can visualize in Fig. 17, it is only when a bfl flip occurs during a bang-bang period that the net influence of the bfl felt by the qubit is nonzero, and the qubit thus suffers some random deviation from its trajectory in the noiseless case. Taken together, these random deviations constitute a random walk around the noiseless trajectory. While this walk is actually continuous, it

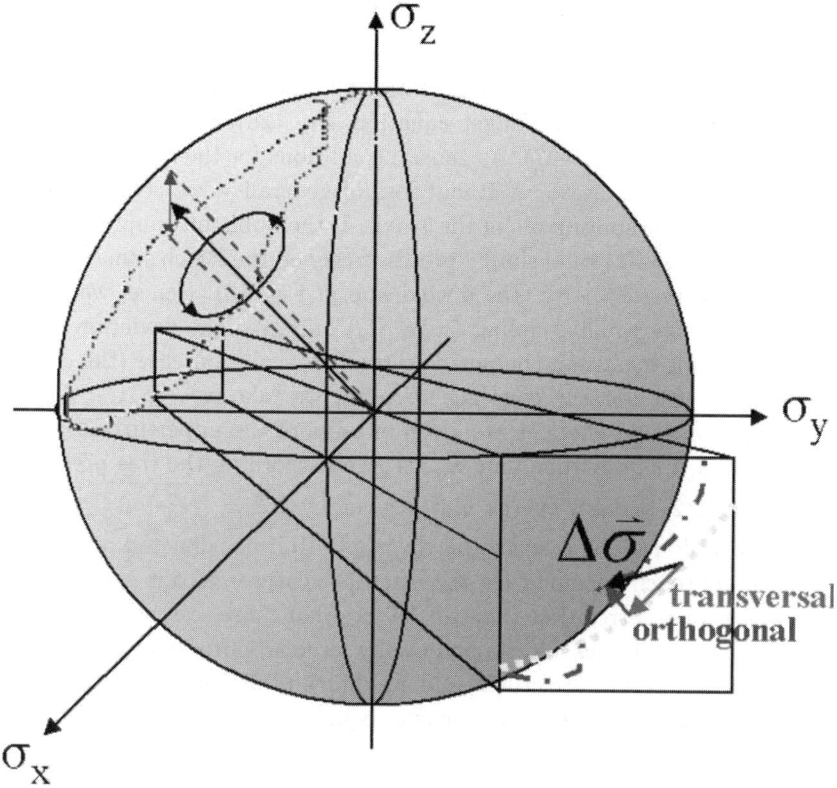

Fig. 18. Schematic plot of noisy qubit evolution generated by Poissonian telegraph noise. The resulting random walk (dot-dashed line) on the Bloch sphere is comprised both of deviations $\Delta\sigma_{\rm deph}$ in parallel to the free precession trajectory (dotted line), which correspond to dephasing, and deviations $\Delta\sigma_{\rm rel}$ perpendicular to it, which correspond to relaxation/excitation.

can be modelled as a discrete walk with steps that are randomly distributed in time, one step for each bfl flip (see *e.g.* [184]). The average step length is essentially the product of the noise coupling strength α and the mean time the bfl in its present state can influence the qubit. Without bang-bang control, this mean influence time is $\tau_{\rm bfl}$, whereas with bang-bang control, it is reduced to $\tau_{\rm bb}$. Therefore, both with and without bang-bang control, the random walk has the same time distribution of steps, but with bang-bang control the step size can be significantly reduced roughly by a factor of the ratio of time scales $\tau_{\rm bb}/\tau_{\rm bfl}$.

7.9. *Random Walk on the Bloch sphere*

Now we study this proposal quantitatively. We simulate these random walks both with and without bang-bang control by integrating both numerically and analytically the Schrödinger equation, Eq. (28), with the stochastic Hamiltonian of Eqs. (25-27). As generic conditions for the qubit dynamics, we choose $\epsilon_{\mathrm{q}} = \Delta_{\mathrm{q}} \equiv \Omega_0$. Without loss of generality, we set the qubit's initial state to be spin-up along the z-axis. If the qubit-bfl coupling α were zero, then the qubit would simply precess freely on the Bloch sphere around the rotation axis $\hat{\sigma}_{\mathrm{x}}^{\mathrm{q}} + \hat{\sigma}_{\mathrm{z}}^{\mathrm{q}}$ (the dotted line in Fig. 18). Hence, we expect for a sufficiently small coupling ($\alpha \ll \Omega_0$) only a slight deviation of the individual time evolution compared to the free evolution case (the dashed line in Fig. 18). For the coupling strength, we take $\alpha = 0.1\Omega_0$. All the following times and energies are given in units of the unperturbed system Hamiltonian, *i.e.*, our time unit τ_{Sys} is given according the free precession time $\pi\tau_{\mathrm{Sys}}/\sqrt{2}$, and our energy unit is given by $\Delta E = \sqrt{\epsilon_{\mathrm{q}}^2 + \Delta_{\mathrm{q}}^2} = \sqrt{2}\Omega_0$. The time scale ratio is taken to be $\tau_{\mathrm{bfl}}/\tau_{\mathrm{bb}} = 10$ if not denoted otherwise.

This approach accounts for the essential features of our specific situation: the long correlation time of the external noise, essentially τ_{bfl}, its non-Gaussian statistics and its potentially large amplitude at low frequencies. These properties are crucial and are difficult, although not impossible, to take into account in standard master equation methods.

7.9.1. *Numerical simulations*

We have numerically integrated Eq. (28) and averaged the deviations of the random walk evolution from the unperturbed trajectories for times up to $100\tau_{\mathrm{Sys}}$ over $N = 10^3$ realizations. Larger simulations have proven that convergence is already sufficient at this stage. We shall examine the root-mean-square (rms) deviations of this ensemble at given time points

$$\Delta\vec{\sigma}_{\mathrm{rms}}(t) = \sqrt{\frac{1}{N}\sum_{j=1}^{N}\left(\vec{\sigma}_j^{\mathrm{q}}(t) - \vec{\sigma}_{\mathrm{noisy},j}^{\mathrm{q}}(t)\right)^2} \qquad (29)$$

with and without bang-bang control. In other approaches, such as those based on master equations, one separates dephasing and relaxation. Both are contained here in Eq. (29). We shall point out notable differences between these two channels. The deviation as a function of time is plotted in Fig. 19.

The total deviations on intermediate time scales are suppressed by a

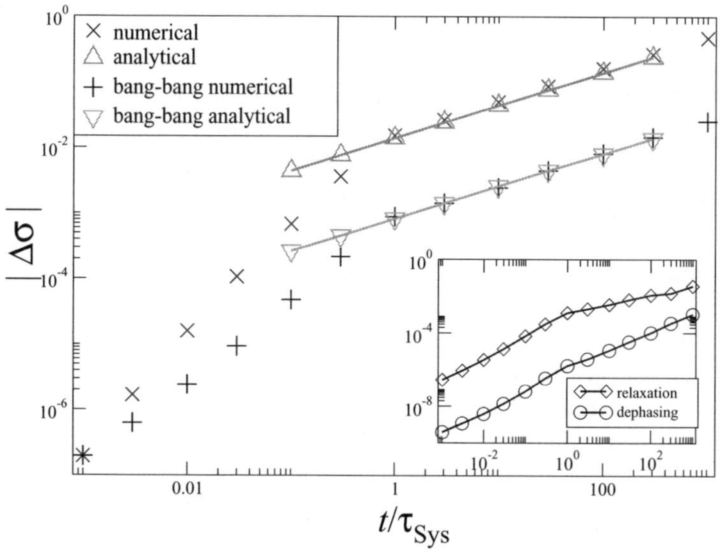

Fig. 19. Time evolution of the rms deviations for bfl-induced random walks with and without bang-bang control at a coupling constant $\alpha = 0.1$ and a typical flipping time scale $\tau_{\rm bfl} = 10^{-2}\tau_{\rm Sys}$. The separation between two bang-bang pulses is $\tau_{\rm bb} = 10^{-3}\tau_{\rm Sys}$. The straight lines are square-root fits of the analytical derived random walk model variances (plotted as triangles). Inset: Components of the deviations from the free precession trajectory that are parallel to it (dephasing) and perpendicular to it (relaxation/excitation) with bang-bang control.

ratio on the order of 10. A detailed numerical analysis shows that *without* bang-bang suppression, the deviations parallel to the free precession trajectory (which correspond to dephasing) are of similar size to those perpendicular to free precession (which correspond to relaxation/excitation). In contrast, with bang-bang control, dephasing is almost totally absent as one can see in the inset of Fig. 19.

The main double-logarithmic plot of Fig. 19 shows that on short time scales ($t \simeq 0.1\tau_{\rm Sys}$, which corresponds to $\simeq 10$ random walk steps), deviations increase almost linearly in time. It is not until times on the order of $\tau_{\rm Sys}$ that the noise-induced deviations start to behave as typical classical random walks, increasing as a square-root in time.

7.9.2. *Analytical random walk models*

We now develop analytical random walk models for our system. The random walk on the Bloch sphere is in general two-dimensional, consisting of both parallel and perpendicular deviations to the free evolution trajectory. Bang-bang control, as was seen in the above numerical results and as will be seen in the following analytical results, essentially reduces the random walk to one-dimension as only the perpendicular deviations remain significant. In the following, we restrict ourselves to the long-time (many random walk steps) regime.

We first calculate for both cases the probability distributions of the deviations after one bfl flip ("one-step deviations" in terms of the discrete random walk). The fluctuation of the period between $\tau_{\text{per}}^{\pm}$ leads to dephasing, which can be evaluated at $\alpha \ll \epsilon_{\text{q}}, \Delta_{\text{q}}$ to

$$\Delta\vec{\sigma}_{\text{deph}}^{\text{bfl}} = 2\pi\cos\phi\left(\frac{1}{\tau_{\text{per}}^{\pm}} - \frac{1}{\tau_{\text{per}}}\right)\tau_{\text{bfl}} \simeq \pm 2\frac{\Delta_{\text{q}}\epsilon_{\text{q}}}{\Delta_{\text{q}}^2+\epsilon_{\text{q}}^2}\alpha\tau_{\text{bfl}}. \tag{30}$$

For the relaxation/excitation effect of the noise, one has to use the projection of the perturbation orthogonal to the free axis.

In total, using $\tau_{\text{per}}^{\pm} \simeq \tau_{\text{per}}$ to first order in α, we find

$$\Delta\vec{\sigma}_{\text{rel}}^{\text{bfl}} = 2\pi\cos\phi\sin\eta\frac{1}{\sqrt{2}}\cos\phi\frac{\tau_{\text{bfl}}}{\tau_{\text{per}}^{\pm}} \simeq \sqrt{2}\frac{\Delta_{\text{q}}^3}{(\epsilon_{\text{q}}^2+\Delta_{\text{q}}^2)^{3/2}}\alpha\tau_{\text{bfl}}. \tag{31}$$

The derivation of the maximal one-step deviation for the bang-bang controlled situation has to be handled differently. The deviation resulting from a bfl flip during a bang-bang pulse period is maximal if the step happens exactly at the moment of the second qubit spin-flip (*i.e.*, in the middle of the bang-bang cycle). When this happens, the refocusing evolution has in its first half a drift, for example, to the "right" (compare Fig. 21) and in the last half an equal aberration.

We average the maximal one-step deviation over one precession period in the usual rms manner to obtain

$$\begin{aligned}\left\langle\Delta\vec{\sigma}_{\text{max}}^{\text{bb}}\right\rangle &= \sqrt{\frac{1}{2\pi}\int_0^{2\pi}\sin^2\chi 4\alpha^2\tau_{\text{bb}}^2\,d\chi} \\ &= \sqrt{2}\alpha\tau_{\text{bb}}.\end{aligned} \tag{32}$$

Obviously, this variance only contributes to relaxation.

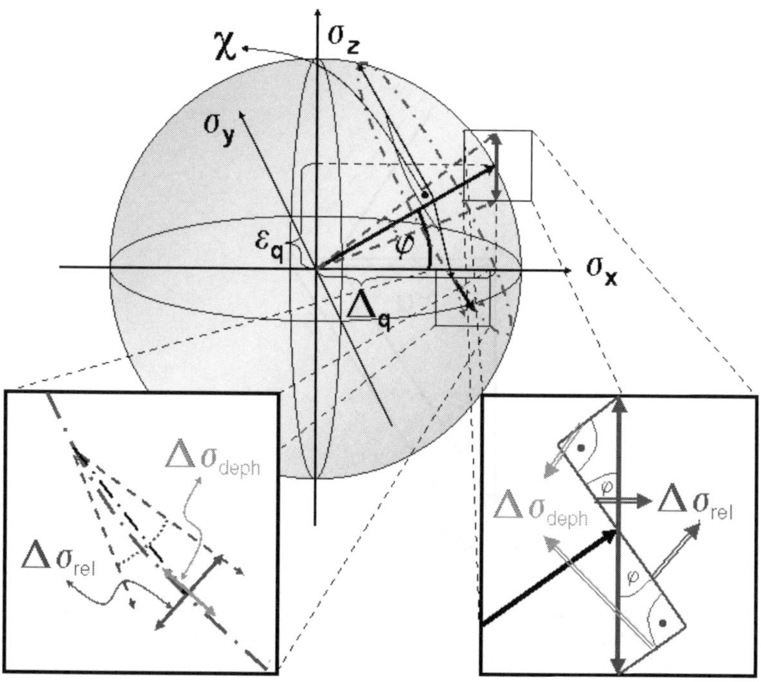

Fig. 20. Plot of a typical one-step deviation from the unperturbed qubit trajectory with generic values for ϵ_{q} and Δ_{q}. The fractions of the bfl fluctuations in $\hat{\sigma}_{\mathrm{z}}$-direction have to be distinguished with respect to their effects on the qubit: those that yield dephasing deviations that are parallel to the free precession trajectory (proportional to $\sin\phi$) versus relaxation/excitation deviations that are perpendicular (proportional to $\sin\eta$). Both parts are additionally domineered by a factor of $\cos\phi$ due to the diminished radius of the trajectory starting from the initial state $\sigma_z = +1$. The impact of the relaxation/excitation generating part is furthermore depending on $\cos\phi$ as well as $\sin\chi$, the azimuth angle of the qubits present position.

From the long-time limit of our analytical random walk distribution, we find for their variances in real space representation

$$\Delta\sigma_{\mathrm{bfl}}(N_{\mathrm{bfl}}) = \sqrt{N_{\mathrm{bfl}}}\beta = \sqrt{N_{\mathrm{bfl}}}\frac{\sqrt{5}}{2}\alpha\tau_{\mathrm{bfl}} \tag{33}$$

for the case without bang-bang control and

$$\Delta\sigma_{\mathrm{bb}}(N_{\mathrm{bfl}}) = \frac{\sqrt{N_{\mathrm{bfl}}}}{2}\gamma = \sqrt{\frac{N_{\mathrm{bfl}}}{2}}\alpha\tau_{\mathrm{bb}} \tag{34}$$

for the case with it. In the large-N_{bfl} limit, this model shows excellent agreement with the numerical simulations.

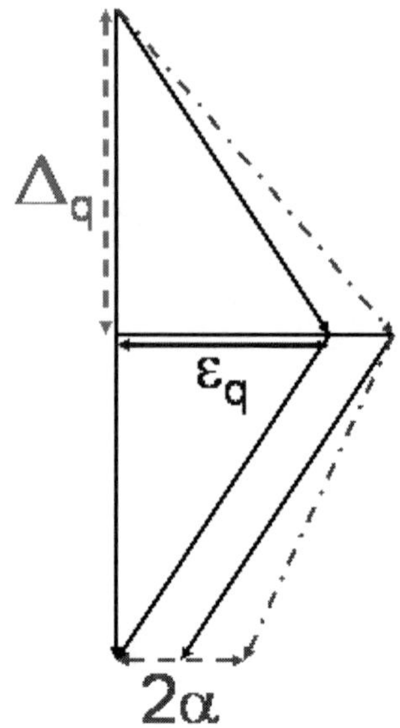

Fig. 21. Sketch of a maximal one-step deviation during a bang-bang modulated cycle, which appears if the bfl state flips precisely at the intermediate bang-bang pulse time. The dephasing part of deviation evidently averages out, while a relaxating aberrance arise proportional to the noise-coupling constant α.

7.9.3. *Bang-bang control working as a high-pass filter*

In order to measure the degree of noise suppression due to bang-bang control, we define the suppression factor $\mathcal{S}_{t_0}$ as follows for a given evolution time t_0

$$\mathcal{S}_{t_0}(\tau_{\mathrm{bfl}}/\tau_{\mathrm{bb}}) \equiv \frac{\Delta\vec{\sigma}^{\mathrm{bfl}}_{\mathrm{rms}}(t_0)}{\Delta\vec{\sigma}^{\mathrm{bb}}_{\mathrm{rms}}(t_0)}. \tag{35}$$

We now systematically study the dependence of $\mathcal{S}_{t_0}$ on $\tau_{\mathrm{bfl}}/\tau_{\mathrm{bb}}$ for a constant mean bfl switching rate $\tau_{\mathrm{bfl}} = 10^{-2}\tau_{\mathrm{sys}}$ at a fixed evolution time $t_0 = \tau_{\mathrm{sys}}$. The numerical data in Fig. 22 show that the suppression efficiency is linear in the bang-bang repetition rate, $S_{\tau_{\mathrm{sys}}} = \mu\tau_{\mathrm{bfl}}/\tau_{\mathrm{bb}}$. The numerically derived value of the coefficient, $\mu_{\mathrm{numerical}} \approx 1.679$, is in excellent agreement with the analytical result $\mu_{\mathrm{analytical}} = \sqrt{5/2} \simeq 1.581$ from our saddle point approximation, Eqs. (33) and (34).

We have investigated the qubit errors that arise from the noise gener-

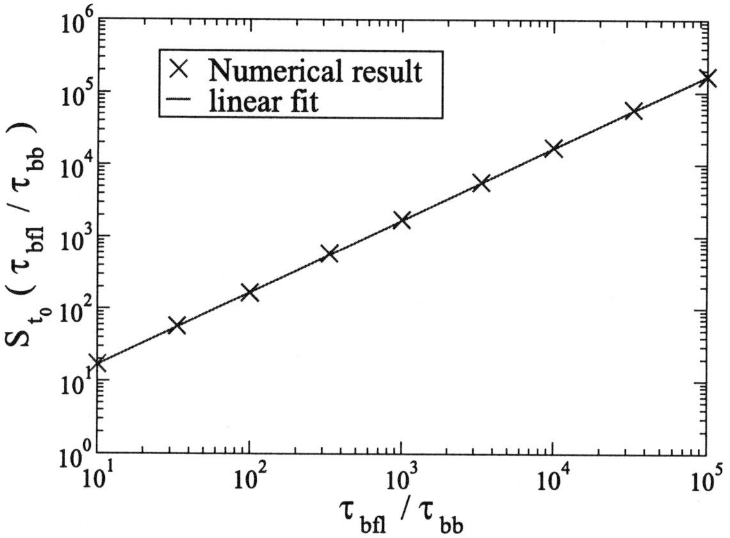

Fig. 22. The suppression factor $\mathcal{S}_{t_0}(\tau_{\rm bfl}/\tau_{\rm bb}) = \frac{\Delta\vec{\sigma}^{\rm bfl}_{\rm rms}(t_0)}{\Delta\vec{\sigma}^{\rm bb}_{\rm rms}(t_0)}$ evaluated for $t_0 = \tau_{\rm Sys}$ as a function of the ratio of the mean switching time $\tau_{\rm bfl}$ and the bang-bang pulse separation $\tau_{\rm bb}$.

ated by a *single* bistable fluctuator (bfl) in its semiclassical limit, where it behaves as a telegraph noise source. We numerically integrated a corresponding stochastic Schrödinger equation, Eq. (28), as well as analytically solved (in the long-time limit) appropriate random walk models. As a characteristic measure of the resulting dephasing and relaxation effects, we used the rms deviation of noisy evolutions compared to noiseless ones. To suppress the effects of this noise, we presented a bang-bang pulse sequence analogous to the familiar spin-echo method. We claimed this pulse sequence to be capable of refocusing most of the bfl-noise induced aberrations. Both in the case without bang-bang control and the case with it, there was excellent agreement between our numerical and analytical results on the relevant intermediate time scales (*i.e.*, times after a short initial phase where deviations grow linearly instead of as a square-root in time, but before the qubit becomes totally decohered).

Meanwhile, several other extensions of Ref. [185,183] have been proposed by other research groups. Ref. [171] includes a larger number of fluctuators, described as semiclassical noise sources, but restricts itself to a single spin-echo cycle. Ref. [170] analyzes extensively the importance of higher, non-Gaussian cumulants and memory effects and arrives at a number of analytical results, but it does not treat the option of refocusing. Ref. [186] treats a full microscopic model and compares different variations of the bang-bang pulse sequence. Ref. [187] also treats a full microscopic model with potentially many fluctuators using a Lindblad-type approach and covers a wide range of ratios between the fluctuator and bang-bang pulse time scales. One of its main conclusions is that a Zeno effect is found in a parameter regime not covered by our work. Note that all of these other extensions of our work treat only the case of ideal bang-bang pulses.

8. Coupled qubits and beyond

To implement a quantum algorithm, one must be able to entangle multiple qubits, so that an interaction term is required in the Hamiltonian describing a two qubit system. For two superconducting flux qubits, the natural interaction is between the magnetic fluxes. Placing the two qubits in proximity provides a permanent coupling through their mutual inductance [188]. Pulse sequences for generating entanglement have been derived for several superconducting qubits with fixed interaction energies [112,189]. However, entangling operations can be much more efficient if the interaction can be varied and, ideally, turned off during parts of the manipulation. A variable coupling scheme for charge-based superconducting qubits with a bipolar interaction has been suggested recently [190]. For flux qubits, while switchable couplings have been proposed previously [73,191], these approaches do not enable one to turn off the coupling entirely and require separate coupling and flux readout devices.

As a new device, we propose a new coupling scheme for flux qubits in which the interaction is adjusted by changing a relatively small current. For suitable device parameters the sign of the coupling can also be changed, thus making it possible to null out the direct interaction between the flux qubits. Furthermore, the same device can be used both to vary the coupling and to read out the flux states of the qubits. We show explicitly how this variable qubit coupling can be combined with microwave pulses to perform the quantum Controlled-NOT (CNOT) logic gate. Using microwave pulses also for arbitrary single-qubit operations, this scheme provides all the necessary ingredients for implementation of scalable univeral quantum logic.

The coupling is mediated by the circulating current J in a dc Superconducting QUantum Interference Device (SQUID), in the zero voltage state, which is coupled to each of two qubits through an identical mutual inductance M_{qs} [Fig. 23(a)]. A variation in the flux applied to the SQUID, Φ_s, changes J [Fig. 23(b)]. The response is governed by the screening parameter $\beta_L \equiv 2LI_0/\Phi_0$ and the bias current I_b, where $I_b < I_c(\Phi_s)$, the critical current for which the SQUID switches out of the zero voltage state at $T = 0$ in the absence of quantum tunneling. In flux qubit experiments [75], the flux state is determined by a dc SQUID to which fast pulses of I_b are applied to measure $I_c(\Phi_s, T)$. Thus, existing technology allows I_b to be varied rapidly, and a single dc SQUID can be used both to measure the two qubits and to couple them together controllably.

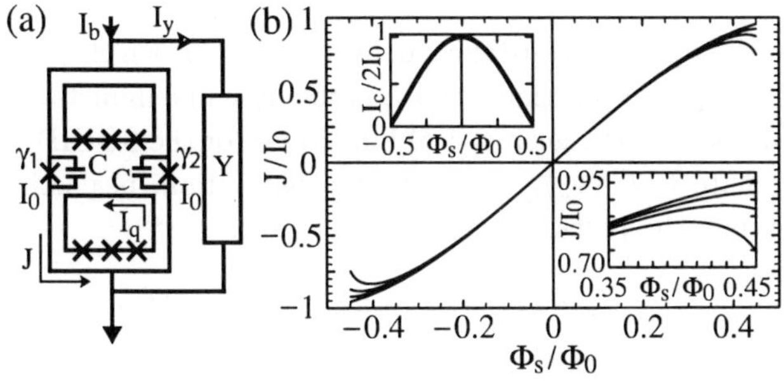

Fig. 23. (a) SQUID-based coupling scheme. The admittance Y represents the SQUID bias circuitry. (b) Response of SQUID circulating current J to applied flux Φ_s for $\beta_L = 0.092$ and $I_b/I_c(0.45\Phi_0) = 0, 0.4, 0.6, 0.85$ (top to bottom). Lower right inset shows $J(\Phi_s)$ for same values of I_b near $\Phi_s = 0.45\Phi_0$. Upper left inset shows I_c versus Φ_s.

The energy biases ϵ_i^0 are determined by the flux bias of each qubit relative to $\Phi_0/2$. The tunnel frequencies δ_i/h are fixed by the device parameters, and are typically a few GHz. For two flux qubits, arranged so that a flux change in one qubit alters the flux in the other, the coupled-qubit Hamiltonian describing the dynamics in the complex 4-dimensional Hilbert space becomes

$$\mathcal{H} = \mathcal{H}_1 \otimes I^{(2)} + I^{(1)} \otimes \mathcal{H}_2 - (K/2)\sigma_z^{(1)} \otimes \sigma_z^{(2)}, \tag{36}$$

where $I^{(i)}$ is the identity matrix for qubit i and K characterizes the coupling energy. For $K < 0$, the minimum energy configuration corresponds to antiparallel fluxes. For two flux qubits coupled through a mutual inductance M_{qq}, the interaction energy is fixed at $K_0 = -2M_{qq}\left|I_q^{(1)}\right|\left|I_q^{(2)}\right|$.

For the configuration of Fig. 23(a), in addition to the direct coupling, K_0, the qubits interact by changing the current J in the SQUID. The response of J to a flux change depends strongly on I_b [Fig. 23(b)]. When $I_q^{(2)}$ switches direction, the flux coupled to the SQUID, $\Delta\Phi_s^{(2)}$, induces a change ΔJ in the circulating current in the SQUID, and alters the flux coupled from the SQUID to qubit 1. The corresponding coupling is

$$K_s = I_q^{(1)}\Delta\Phi_q^{(1)} = -2M_{qs}^2 \left|I_q^{(1)}\right| \left|I_q^{(2)}\right| \mathrm{Re}\,(\partial J/\partial\Phi_s)_{I_b}\,. \tag{37}$$

The transfer function, $(\partial J/\partial\Phi_s)_{I_b}$, is related to the dynamic impedance, Z, of the SQUID via [192]

$$\partial J/\partial\Phi_s = i\omega/Z = 1/\mathcal{L} + i\omega/R, \tag{38}$$

where R is the dynamic resistance, determined by Y which dominates any loss in the Josephson junctions, and $\mathcal{L}$ is the dynamic inductance which, in general, differs from the geometrical inductance of the SQUID, L.

We evaluate $(\partial J/\partial\Phi_s)_{I_b}$ by current conservation, neglecting currents flowing through the junction resistances:

$$I_b = I_y + 2I_0\cos\Delta\gamma\sin\bar{\gamma} - 2C(\Phi_0/2\pi)\ddot{\bar{\gamma}}, \tag{39}$$

$$J = I_0\cos\bar{\gamma}\sin\Delta\gamma - C(\Phi_0/2\pi)\Delta\ddot{\gamma}. \tag{40}$$

Here, I_y is the current flowing through the admittance $Y(\omega)$ [Fig. 23(a)], and I_0 and C are the critical current and capacitance of each SQUID junction. The phase variables are related to the phases across each junction, γ_1 and γ_2, as $\Delta\gamma = (\gamma_1-\gamma_2)/2$ and $\bar{\gamma} = (\gamma_1+\gamma_2)/2$. The phases are constrained by $d\Delta\gamma = (\pi/\Phi_0)(d\Phi_s - L dJ)$.

The expression for K_s in terms of $\mathrm{Re}\,(\partial J/\partial\Phi_s)_{I_b}$ (Eq. 37) requires the qubit frequencies to be much lower than the characteristic frequencies of the SQUID. This condition is satisfied by our choice of device parameters, and also ensures that the SQUID stays in its ground state during qubit entangling operations. Furthermore, it is a reasonable approximation to take the $\omega = 0$ limit of $\mathrm{Re}\,(\partial J/\partial\Phi_s)_{I_b}$ to calculate K_s, so that we can solve Eqs. (39) and (40) numerically to obtain the working point; for the moment we assume $Y(0) = 0$. For the small deviations determining K_s, we linearize Eqs. (39) and (40) and solve for the real part of the transfer function in the low-frequency limit:

$$\mathrm{Re}\left(\frac{\partial J}{\partial\Phi_s}\right)_{I_b} = \frac{1}{2L_j}\frac{1-\tan^2\Delta\gamma\tan^2\bar{\gamma}}{1+\frac{L}{2L_j}(1-\tan^2\Delta\gamma\tan^2\bar{\gamma})}. \tag{41}$$

Here, we have introduced the Josephson inductance for one junction, $L_j = \Phi_0/2\pi I_0 \cos\Delta\gamma \cos\bar{\gamma}$. For $\beta_L \gg 1$, Eq. (41) approaches $1/L$, while for $\beta_L \ll 1$,

$$\mathrm{Re}\,(\partial J/\partial\Phi_s)_{I_b} = (1/2L_j)(1 - \tan^2\Delta\gamma\tan^2\bar{\gamma}). \tag{42}$$

We see that $\mathrm{Re}\,(\partial J/\partial\Phi_s)_{I_b}$ becomes negative for sufficiently high values of I_b and Φ_s, which increase $\bar{\gamma}$ and $\Delta\gamma$.

We choose the experimentally-accessible SQUID parameters $L = 200$ pH, $C = 5$ fF, and $I_0 = 0.48$ μA, for which $\beta_L = 0.092$. The qubits are characterized by $I_q^{(1)} = I_q^{(2)} = 0.46$ μA, $M_{qs} = 33$ pH, and $M_{qq} = 0.25$ pH, yielding $K_0/h = -0.16$ GHz. Choosing $\Phi_s = 0.45\Phi_0$, Eqs. (37) and (41) result in a net coupling strength $K/h = (K_0 + K_s)/h$ that is -0.3 GHz when $I_b = 0$, and zero when $I_b/I_c(0.45\Phi_0) = 0.57$ [Fig. 24(a)]. The change in sign of K_s does not occur for all β_L. Figure 24(b) shows the highest achievable value of K_s versus β_L. We have adopted the optimal design at $\beta_L = 0.092$.

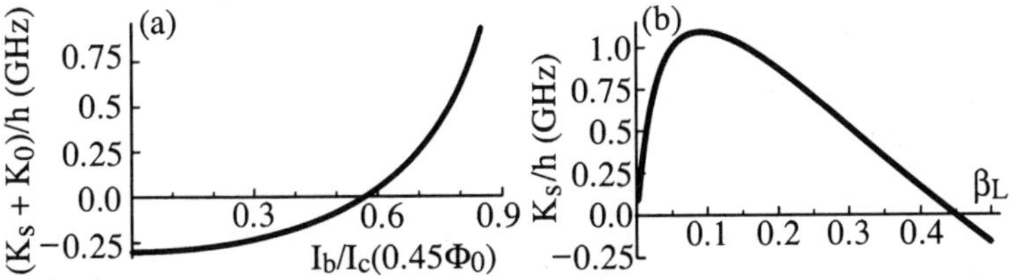

Fig. 24. (a) Variation of K with I_b for $\Phi_s = 0.45\Phi_0$ and device parameters described in text. (b) Highest achievable value of K_s versus β_L evaluated at $I_b = 0.85I_c(0.45\Phi_0)$; I_0 (and hence β_L) is varied for $L = 200$ pH.

We also need to consider crosstalk between the coupling and single-qubit terms in the Hamiltonian. When the coupling is switched, in addition to $\partial J/\partial\Phi_s$ being altered, J also changes, thus shifting the flux biases of the qubits. The calculated change in J as the coupler is switched from $I_b = 0$ to $I_b/I_c(0.45\Phi_0) = 0.57$ produces a change in the flux in each qubit corresponding to an energy shift $\delta\epsilon_1/h = \delta\epsilon_2/h = 1.64$ GHz. In addition, when the qubits are driven by microwaves to produce single-qubit rotations, the microwave flux may also couple to Φ_s. As a result, K is weakly modulated when the coupling would nominally be turned off. A typical microwave drive $\tilde{\epsilon}_i(t)/h$ of amplitude 1 GHz results in a variation of about ± 14 MHz about $K = 0$.

When the bias current is increased to switch off the coupling, the SQUID symmetry is broken and the qubits are coupled to the noise generated by the admittance Y. We estimate the decoherence due to this process using the technique outlined in section 7.3. We obtain as an intermeiate result

$$\mathcal{J}(\omega) = \left(I_q^2 M_{qs}^2/h\right) \mathrm{Im}\left(\partial J/\partial \Phi_s\right)_{I_b}. \tag{43}$$

For the case $Y^{-1} = R$, following the path to the static transfer function Eq. (41) and taking the imaginary part in the low-β_L limit, we obtain $\mathrm{Im}(\partial J/\partial \Phi_s)_{I_b} = -\omega/R = (\omega/4R)\tan^2\Delta\gamma\tan^2\bar{\gamma}$. Thus $J(\omega) = \alpha\omega$, where $\alpha = (M_{qs}^2 I_q^2/4hR)\tan^2\Delta\gamma\tan^2\bar{\gamma}$, and $\alpha(I_b = 0) = 0$. As I_b is increased to change the coupling strength, α increases monotonically. For the parameters described above and for $R = 2.4$ kΩ, when the net coupling is zero $[I_b/I_c(0.45\Phi_0) = 0.57$, Fig. 24(a)], we find $\alpha = 8 \times 10^{-5}$ corresponding to a qubit dephasing time of about 500 ns, one order of magnitude larger than values currently measured in flux qubits [75], however, shorter than the best values with flux echo.

We now show that this configuration implements universal quantum logic efficiently. Any n-qubit quantum operation can be decomposed into combinations of two-qubit entangling gates, for example, CNOT, and single-qubit gates [193]. Single-qubit gates generate local unitary transformations in the complex 2-dimensional subspace for the corresponding individual qubit, while the two-qubit gates correspond to unitary transformations in the 4-dimensional Hilbert space. Two-qubit gates which cannot be decomposed into a product of single-qubit gates are said to be nonlocal, and may lead to entanglement between the two qubits [194]. Since we can adjust the qubit coupling K to zero, we can readily implement single-qubit gates with microwave pulses as described below.

To implement the nonlocal two-qubit CNOT gate, we use the concept of local equivalence: the two-qubit gates U_1 and U_2 are locally equivalent if $U_1 = k_1 U_2 k_2$, where k_1 and k_2 are local two-qubit gates which are combinations of single-qubit gates applied simultaneously. These unitary transformations on the two single-qubit subspaces transform the gate U_2 into U_1. The local gate which precedes U_2, k_2, is given by $k_{21} \otimes k_{22}$, where $k_{21(22)}$ is a single-qubit gate for qubit 1(2), while the local gate which follows U_2, k_1, is $k_{11} \otimes k_{12}$, where $k_{11(12)}$ is a single-qubit gate for qubit 1(2) [195]. Our strategy is to find efficient implementation of a nonlocal quantum gate U_2 that differs only by local gates, k_1 and k_2, from CNOT, using the methods in [194], and the computational basis, in which the SQUID measures the projection of each qubit state vector onto the z-axis.

The local equivalence classes of two-qubit operations have been shown

[194] to be in one-to-one correspondence with points in a tetrahedron, the Weyl chamber. In this geometric representation, any two-qubit operation is associated with the point $[c_1, c_2, c_3]$, where CNOT corresponds to $[\pi/2, 0, 0]$. Furthermore, the nonlocal two-qubit gates generated by a Hamiltonian acting for time t can be mapped to a trajectory in this space [194]. If K is increased instantaneously to a constant value, the trajectory generated by Eq. (36) is well described by the following periodic curve

$$[c_1, c_2, c_3] = [Kvt/\hbar, p\,|\sin \omega t|\,, p\,|\sin \omega t|]\,. \tag{44}$$

Here, p is a function of the system parameters, $v = \epsilon_1^0 \epsilon_2^0 / \Delta E_1 \Delta E_2$, and $\omega = (\Delta E_1 - \Delta E_2)/2\hbar$, where $\Delta E_i = [(\epsilon_i^0)^2 + \delta_i^2]^{1/2}$ is the single-qubit energy level splitting. Independently of p, this trajectory reaches $[\pi/2, 0, 0]$ in a time $t_K = n\pi/\omega$ when the coupling strength is tuned to $K = \hbar\omega/2nv$, with n a nonzero integer.

While this analytic solution contains the essential physics, it is an approximation and does not include vital experimental features, in particular, crosstalk and the finite rise time of the bias current pulse. To improve the accuracy, we perform a numerical optimization using Eq. (44) as a starting point, then add these corrections. We use tunnel frequencies $\delta_1/h = 5\,\text{GHz}$ and $\delta_2/h = 3\,\text{GHz}$, and include the shifts of the single-qubit energy biases due to the crosstalk with K_s in Eq. (44) by adding a shift $\delta\epsilon_i$ proportional to K. We account for the rise and fall times of the current pulse by using pulse edges with 90% widths of 0.5 ns [see $K(t)$ in Fig. 25]. We numerically optimize the variable parameters to minimize the Euclidean distance between the actual achieved gate and the desired Weyl chamber target CNOT gate. We find $K/h = -0.30\,\text{GHz}$, $\epsilon_1^0/h = 8.06\,\text{GHz}$, $\epsilon_2^0/h = 2.03\,\text{GHz}$, and $t_K = 8.74\,\text{ns}$; t_K is the time during which the qubit coupling is turned on.

As outlined above, to achieve a true CNOT gate we still have to determine the pulse sequences which implement the requisite local gates that take this Weyl chamber target U_2 to CNOT in the computational basis. Local gates may be implemented by applying microwave radiation, $\tilde{\epsilon}_i(t)$, which couples to $\sigma_z^{(i)}$, and is at or near resonance with the single-qubit energy level splitting ΔE_i. We note that the single-qubit Hamiltonian driven by a resonant oscillating microwave field does not permit one to use standard NMR pulses, since the static and oscillating fields are not perpendicular, but rather are canted by an angle $\tan^{-1}(\delta_i/\epsilon_i^0)$. To simplify the pulse sequence, we keep $\epsilon_{1,2}^0$ constant at the values used for the non-local gate generation. This imposes an additional constraint on the local gates: to generate a local two-qubit gate $k_1 = k_{11} \otimes k_{12}$, the two single-qubit gates

k_{11} and k_{12} must be simultaneous and of equal duration. We satisfy this constraint by making the microwave pulse addressing one qubit resonant and that addressing the other slightly off-resonance. Using this offset and the relative amplitude and phase of the two microwave pulses as variables, we can achieve two different single-qubit gates simultaneously, leading to our required local two-qubit gate.

The resulting pulse sequences for K and $\tilde{\epsilon}_{1,2}$ are shown in Fig. 25. The gate has a maximum deviation from CNOT in the computational basis of 1.6% in any matrix element. This error arises predominantly from the cross-coupling of the microwave signals for the two qubits and the weak modulation of the $K = 0$ state of the coupler during the single-qubit microwave manipulations. While small, this error could be reduced further by performing the numerical optimization with higher precision or by coupling the microwave flux selectively to each of the qubits and not to the SQUID. The total elapsed time of 29.35 ns is comparable to measured dephasing times in a single flux qubit [75].

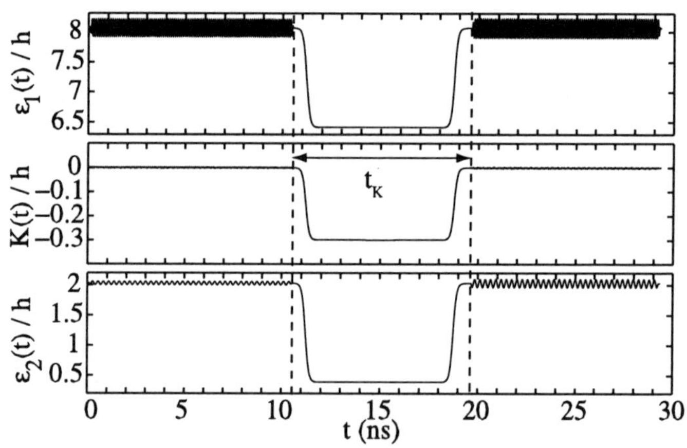

Fig. 25. Pulse sequence for implementing CNOT gate. Energy scales in GHz. Total single-qubit energy bias $\epsilon_i(t) = \epsilon_i^0 + \tilde{\epsilon}_i(t) + \delta\epsilon_i(t)$, where microwave pulses $\tilde{\epsilon}_{1,2}(t)$ produce single-qubit rotations in the decoupled configuration; crosstalk modulation of $K(t)$ is shown (see text). The bias current is pulsed to turn on the interaction in the central region.

We have shown in the preceeding paragraphs that the inverse dynamic inductance of a dc SQUID with low β_L in the zero-voltage state can be varied by pulsing the bias current. This technique provides a variable-strength

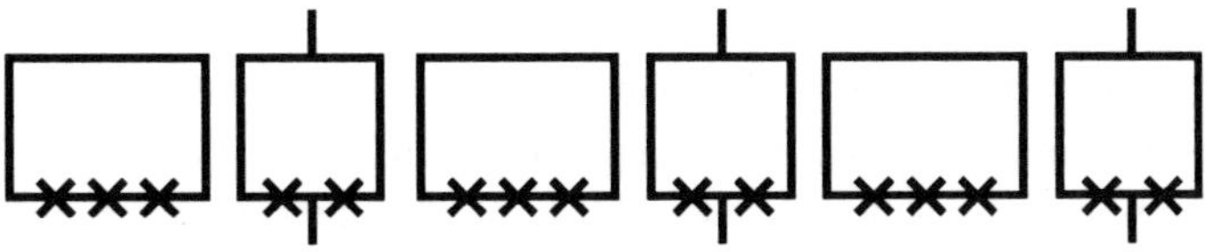

Fig. 26. Chain of flux qubits with intervening dc SQUIDs arranged to provide both variable nearest neighbor coupling and qubit readout.

interaction K_s between flux qubits coupled to the SQUID, and enables cancellation of the direct mutual inductive coupling K_0 between the qubits so that the net coupling K can be switched from a substantial value to zero. By steering a nonlocal gate trajectory and combining it with local gates composed of simultaneous single-qubit rotations driven by resonant and off-resonant microwave pulses, we have shown that a simple pulse sequence containing a single switching of the flux coupling for fixed static flux biases results in a CNOT gate and full entanglement of two flux qubits on a timescale comparable to measured decoherence times for flux qubits. Furthermore, the same SQUID can be used to determine the flux state of the qubits. This approach should be readily scalable to larger numbers of qubits, as, for example, in Fig. 26.

A wide range of other coupling schemes have been proposed and a few of them realized, which we will briefly review here. We will indicate a coupling Hamiltonian of the form $K\sigma_a \otimes \sigma_b$ simply as AB.

A constant ZZ coupling using a flux transformer has been demonstrated for flux qubits [188]. This is expcted to be possible up to strong coupling [196]. This transformer can in principle be made tunable by including a DC-SQUID loop [73,74] or another superconducting switch [197] into the loop. Alternatively, the conjugate variable, charge, can be used in capacitive coupling, leading to a non-tunable $XX + YY$ interaction [198]. All of these proposals need one physical coupling element per interacting pair, so they will conceivably be of nearest-neighbor type.

For phase qubits, the coupling is not straightforwardly put into a Pauli matrix representation and the coupling matrix is generally bias dependent [199] and also depends on whether the qubits are encoded into adiabatic eigenstates or a global basis; however, the resulting interaction certainly has entangling power [189]. A spectroscopic hint on two-qubit entanglement [200] and three qubit interaction [201] has been observed. Very recently, coherent manipulations have been achieved in this system in a setup which allows for simultaneous measurement [113].

In the charge qubit case, a resonator coil between all qubits has been proposed for a YY interaction [30], which is tunable and of arbitrary range but does not appear to be compatible with parallel operation. Note, that a large coil usually provides strong coupling to external noise, but it can in principle be replaced by a large classical Josephson junction using the kinetic inductance [202–204]. Using the screening currents inside loop-shaped charge qubits, a tunable XX nearest-neighbor interaction can be implemented [205]. More easily, one can implement a capacitive interaction, which leads to ZZ. This is typically constant; however, using the capacitance of additional Josephson junctions it can be made tunable [190]. Coherent charge dynamics and conditional operations with constant capacitive interaction have been demonstrated [79,112]. A tunable nearest-neighbor $XX + YY$ coupling is achieved by connection with SQUIDs; however, the interaction strength is rather low [206]. Charge qubits can profit from a cavity, mediating an XX-interaction [207] and nonclassical radiation [208].

The interaction between quantronia can be implemented using a capacitor and is expected to be of $XX + YY$ structure.

This variety of possible interactions opens the question, which type of interactions are efficient and about the importance of tunability, long range, and parallel operation. Clearly, tunability is not needed for the demonstration of gates on a few qubits similar to NMR [209,10,112,210]; however, with an increasing number of qubits, controls are supposed to become more and more complex. In fact, the in situ tunability of interactions is one of the main advantages of superconducting qubits and shouldnot be given up. On the other hand, long range interactions are less crucial as algorithms operate efficiently even on linear nearest neighbor arrays [211]. It is on the other hand absolutely crucial to operate in parallel.

9. Summary

Let's take an overview of the present status of superconducting Josephson qubits from a view point of the theme of this conference ; "Are the DiVincenzo criteria satisfied by today's experiments ?"

(1) A scalable physical system of well-characterized qubits

Employing the very advanced micro and nano fabrication technology from the semiconductor industry, superconducting qubits have strong potential advantage in scalability towards the circuit level, including tunable coupling switches.

(2) The ability to initialize the state of the qubits to a simple fiducial state

The requirement of proper initialization can be satisfied if superconducting qubit devices are operated at low enough temperature ~20 mK, which is also far below the critical temperature of aluminum (~1.2 K), niobium (~9 K) or other superconducting materials, such that the ground state is occupied with probability nearly 1. The qubit state is protected from rapid energy relaxation once it is cooled below the critical temperature, because it is disconnected from the phonon bath and there is an energy gap in the quasi-particle density of states.

(3) Long (relative) decoherence times, much longer than the gate-operation time

A typical decoherence time of a present superconducting Josephson qubit is $T_2^* \approx 0.5\mu s$ during free induction decay[81,97,212] and it reaches $T_2 \approx 4\mu s$ under the echo pulse technique[91]. Whereas a typical time for single qubit gate operation, e.g a π-rotation pulse, is an order of 0.1 ns under strong driving. The present few ns π-rotation time via the conditional side-band transition of qubit LC-resonator coupled system[114,118] is ready to improve as a design enabling strong driving. In order to demonstrate C-NOT gate operation, a series of phase-shifted composite side-band pulses are needed. So, the present typical decoherence time normalized by a single two-qubit gate operation time would be an order of 10^3. This experimental value is a result at the optimal operating point, which is a commonly used strategy to decouple qubit to the first order from outside "charge" and "flux" noise[81].

(4) A universal set of quantum gates

An arbitrary rotational gate for a single qubit operation is realized by applying resonant microwave pulses from an on-chip microwave circuit. This is already demonstrated and established in all types of superconducting qubits, i.e., the"charge"[77], "phase"[69,70], "charge-phase"[81,109], and "flux"[75] qubits. In order to achieve noise torelant qubit operation and to achieve two-axis qubit rotation without using slow detuning technique, NMR-like multi pulse sequence control has been demonstrated in a "charge-phase" qubit[95] and also in a "flux"[96] qubit respectively. The antiphase oscillation of the $|01\rangle$ and $|10\rangle$ states in the capacitively coupled Josephson phase two-qubits[113] has been observed. So far, the C-NOT gate operation has been demonstrated only in a "charge" two-qubit system[112].

(5) A qubit-specific measurement capability

Measurement of qubit state can be done, for example, using an extremely sensitive magnetic flux quantum detector namely a SQUID which is placed adjacent to the qubit. Single-shot readout has been achieved in a "flux" qubit for the first time in the slow ramping readout method[213,105] and recently in a "charge" qubit using an S-SET pulsed readout technique[107] and very recently in a "charge-phase" qubit coupled to a transmittion line resonator using reflection wave phase detection[109]. Furthermore, the state of the qubit can be measured by dispersive technique that probes the second derivative of the state energy with respect to qubit bias parameters. The phase degree of freedom is used to perform inductive readout[111] using Josephson bifurcation amplification[110,212]. This novel readout projects the state of the qubit typically in a few hundred nano second which is roughly two orders of magnitude faster than the conventional switching readout which inevitably includes the dead time due to a large numbers of generated quasiparticles to relax to the ground state. This new readout method has a lot of advantages with an improved signal to noise ratio and contrast. Thus the readout visibility of the coherent oscillation is becoming better approaching the unit visibility.

Certainly, theoretical proposals for meeting all DiVincenzo's criteria have been put forward and no principal obstacle is known. However, a few experimental issues in decoherence, in particular the loss of visibility, are not fully understood today and may prove problematic if they are fundamental.

Acknowledgments

We would like to thank Prof. M. Nakahara for organizing this fascinating conference and Kinki University for supporting it. We would like to thank all co-workers with whom we have obtained the results presented here. FKWs work is supported by ARDA through ARO contract No. P-43385-PH-QC and DFG through SFB 631.

References

1. G. Moore, Electronics **38**, 114 (1965).
2. G. Moore, No Exponential is forever, Keynote Presentation International Solid State Circuits Conference, 2003.
3. I. Co., Expanding Moore's law, Available at ftp://download.intel.com/labs/eml/download/EML_opportunity.pdf (2002).
4. M. Nielsen and I. Chuang, *Quantum Computation and Quantum Information* (Cambridge University Press, Cambridge, UK, 2000).
5. J. Preskill, Quantum computation lecture notes, Available at http://www.theory.caltech.edu/people/preskill/ph219/.
6. D. DiVincenzo, Science **270**, 255 (1995).
7. D. DiVincenzo, Fortschr. Phys. **48**, 771 (2000).
8. R. Hughes, the ARDA Quantum Information Science, and T. Panel, A Quantum Information Science and Technology Roadmap, available at http://qist.lanl.gov.
9. W. Kaminsky, S. Lloyd, and T. Orlando, quant-ph/0403090 (unpublished).
10. L. Vandersypen *et al.*, Nature **414**, 883 (2001).
11. N. Gershenfeld and I. Chuang, Science **275**, 350 (1997).
12. C. Monroe *et al.*, Phys. Rev. Lett. **75**, 4714 (1995).
13. J. Cirac and P. Zoller, Phys. Rev. Lett. **74**, 4091 (1995).
14. Q. Turchette *et al.*, Phys. Rev. Lett. **75**, 4710 (1995).
15. T. Pellizzari, S. Gardiner, J. Cirac, and P. Zoller, Phys. Rev. Lett. **75**, 3788 (1995).
16. E. Knill, R. Laflamme, and G. Milburn, Nature **409**, 46 (2001).
17. H.-J. Briegel *et al.*, Journal of modern optics **47**, 415 (2000).
18. D. Kielpinski, C. Monroe, and D. J. Wineland, Nature **417**, 709 (2002).
19. S. Sze, *Semiconductor devices, physics and technology* (Wiley-VCH, New York, 2001).
20. S. Thompson *et al.*, Intel Technology Journal **6**, (2002).
21. Y. Imry, *Introduction to Mesoscopic Physics* (Oxford University Press, Oxford, 1997).
22. *Mesoscopic electron transport*, No. E345 in *Proceedings of the NATO Advanced Study Institute*, edited by L. Kouwenhoven, G. Schön, and L. Sohn (Kluwer, Dordrecht, 1997).
23. A. Leggett, Progr. Theor. Phys. **Suppl. 69**, 80 (1980).
24. A. Leggett and A. Garg, Phys. Rev. Lett. **54**, 857 (1985).
25. C. Tesche, Phys. Rev. Lett. **64**, 2358 (1990).
26. K. Likharev, Usp. Fiz. Nauk. **139**, 169 (1983).
27. P. Anderson, in *Lectures on the Many-Body Problem*, edited by E. Caianiello (Academic Press, New York, 1964), p. 113.
28. *Single Charge Tunneling*, No. B294 in *NATO ASI Series*, edited by H. Grabert and M. Devoret (Plenum, New York, 1992).

29. V. Bouchiat, D. V. et P. Joyez, D. Esteve, and M. Devoret, Physica Scripta **T76**, 165 (1998).
30. Y. Makhlin, G. Schön, and A. Shnirman, Rev. Mod. Phys. **73**, 357 (2001).
31. B. Kane, Nature **393**, 133 (1998).
32. D. Loss and D. DiVincenzo, Phys. Rev. A **57**, 120 (1998).
33. J. Elzerman *et al.*, Phys. Rev. B **67**, 161308 (2003).
34. R. Clark *et al.*, Los Alamos Science **27**, 284 (2002).
35. L. Vandersypen *et al.*, in *Quantum Computing and Quantum Bits in Mesoscopic Systems* (Kluwer, Dordrecht, 2002), Chap. 22.
36. G. Burkhard, H. Engel, and D. Loss, Fortschr. Physik **48**, 965 (2000).
37. R. Hanson *et al.*, Phys. Rev. Lett. **94**, 196802 (2005).
38. H. K. Onnes, Leiden Comm. **122b**, 124 (1911).
39. J. Bardeen, L. Cooper, and J. Schrieffer, Phys. Rev. **108**, 1175 (1957).
40. M. Tinkham, *Introduction to Superconductivity* (McGraw-Hill, New York, 1996).
41. P. de Gennes, *Superconductivity of metals and alloys* (Benjamin, N.Y., 1966).
42. J. Schrieffer, *Theory of Superconductivity* (Cummings, N.Y., 1988).
43. F. London, *Superfluids* (Wiley and Sons, N.Y., 1954).
44. R. Doll and M. Näbauer, Phys. Rev. Lett. **7**, 51 (1961).
45. L. Tian *et al.*, in *Quantum Mesoscopic Phenomena and Mesoscopic Devices in Microelectronics* (Kluwer, Dordrecht, 2000), Chap. Decoherence of the superconducting persistent current qubit, edited by I. O. Kulik and R. Ellialogulu.
46. L. Ioffe, V. Geshkenbein, C. Helm, and G. Blatter, Phys. Rev. Lett. **93**, 057001 (2004).
47. B. Josephson, Phys. Lett. **1**, 251 (1962).
48. A. Barone and G. Paterno, *Physics and applications of the Josephson effect* (Wiley, N.Y., 1982).
49. K. Likharev, *Dynamics of Josephson junctions and circuits* (Gordon and Breach, New York, 1986).
50. A. Golubov, M. Kupriyanov, and E. Il'ichev, Rev. Mod. Phys. **76**, 411 (2004).
51. R. Feynman, *Lectures in Physics* (Addison Wesley, New York, 1970), Vol. 3.
52. L. Landau and E. Lifshitz, *Statistical Physics 2* (Akademie-Verlag, Berlin, 1979).
53. S. Pereverzev *et al.*, Nature **388**, 449 (1997).
54. K. Sukhatme, Y. Mukharsky, T. Chui, and D. Pearson, Nature **411**, 280 (2001).
55. B. P. Anderson and M. A. Kasevich, Science **282**, 1686 (1998).
56. P. Hadley, private communication.
57. T. Orlando and K. Delin, *Foundations of applied superconductivity* (Prentice Hall, New York, 1991).
58. T. van Duzer and C. Turner, *Principles of superconductive devices and circuits* (Elsevier, Amsterdam, 1999).
59. V. Ambegaokar, U. Eckern, and G. Schön, Phys. Rev. Lett. **48**, 1745 (1982).

60. J. Clarke, in *Applications of Superconductivity* (Kluwer, Dordrecht, 2000), Chap. Low- and High-Tc SQUIDs and Some Applications, p. 1, edited by H. Weinstock.
61. J. Niemeyer, Supercond. Sci. and Technology **13**, 546 (2000).
62. J. von Delft and H. Schoeller, Ann. Phys. (Leipzig) **7**, 225 (1998).
63. A. Zaikin and G. Schön, Phys. Rep. **198**, 237 (1990).
64. H. van der Zant and R. Fazio, Phys. Rep. **355**, 236 (2001).
65. R. Feynman and F. Vernon, Ann. Phys. (N.Y.) **24**, 118 (1963).
66. G. Ingold, in *Quantum transport and dissipation* (Wiley-VCH, Weinheim, 1998), Chap. Dissipative quantum systems.
67. A. Caldeira and A. Leggett, Phys. Rev. Lett. **46**, 211 (1981).
68. A. Caldeira and A. Leggett, Ann. Phys. (NY) **149**, 374 (1983).
69. J. Martinis, S. Nam, J. Aumentado, and C. Urbina, Phys. Rev. Lett **89**, 117907 (2002).
70. Y. Yu *et al.*, Science **286**, 889 (2002).
71. A. Berkley *et al.*, Science **300**, 1548 (2003).
72. J. Friedman *et al.*, Nature **46**, 43 (2000).
73. J. Mooij *et al.*, Science **285**, 1036 (1999).
74. T. P. Orlando *et al.*, Phys. Rev. B **60**, 15398 (1999).
75. I. Chiorescu, Y. Nakamura, C. Harmans, and J. Mooij, Science **299**, 1869 (2003).
76. E. Il'ichev *et al.*, Phys. Rev. Lett. **91**, 097906 (2003).
77. Y. Nakamura, Y. Pashkin, and J. Tsai, Nature **398**, 786 (1999).
78. Y. Nakamura, Y. A. Pashkin, T. Yamamoto, and J. S. Tsai, Phys. Rev. Lett. **88**, 047901 (2002).
79. Y. Pashkin *et al.*, Nature **421**, 823 (2003).
80. K. Lehnert *et al.*, Phys. Rev. Lett. **90**, 027002 (2003).
81. D. Vion *et al.*, Science **296**, 286 (2002).
82. L. B. Ioffe *et al.*, Nature **398**, 679 (1999).
83. A. Blais and A. Zagoskin, Phys. Rev. A **61**, 042308 (2000).
84. L. B. Ioffe *et al.*, Nature **398**, 678 (2002).
85. G. J. Dolan, Appl. Phys. Lett. **31**, 3 (1977).
86. A. Leggett, J. Phys. C **14**, 415 (2002).
87. W. Dür, C. Simon, and J. Cirac, Phys. Rev. Lett. **89**, 210402 (2002).
88. D. Crankshaw and T. Orlando, IEEE Trans. Appls. Supercond. **11**, 1006 (2001).
89. A. van den Brink, Phys. Rev. B **71**, 064503 (2005).
90. G. Burkard *et al.*, Phys. Rev. B **71**, 134504 (2005).
91. P. Bertet *et al.*, cond-mat/0412485 (unpublished).
92. A. Leggett *et al.*, Rev. Mod. Phys. **59**, 1 (1987).
93. Y. Nakamura, Y. Pashkin, and J. Tsai, Phys. Rev. Lett. **87**, 246601 (2001).
94. J. Clarke *et al.*, Science **239**, 992 (1988).
95. E. Collin *et al.*, Phys. Rev. Lett. **93**, 157005 (2004).
96. T. Kutsuzawa *et al.*, Appl. Phys. Lett. **87**, 037501 (2005).
97. K. Harrabi, F. Yoshiwara, Y. Nakamura, and J. S. Tsai, lT1696, poster presentation PB-F-92 in the 24th International Conference on Low Tem-

perature Physics (LT24), Orlando, Florida, USA, (2005). (unpublished).
98. K. Kakuyanagi, S. Saito, K. Semba, and H. Takayanagi, to be presented in the Quantum Information Technology Symposium (QIT13), Sendai, Japan, (2005). (unpublished).
99. R. Simmonds *et al.*, Phys. Rev. Lett. **93**, 077003 (2004).
100. K. B. Cooper *et al.*, Phys. Rev. Lett. **93**, 180401 (2004).
101. T. Bauch *et al.*, Phys. Rev. Lett. **94**, 087003 (2005).
102. K. Inomata *et al.*, Phys. Rev. Lett. **95**, 107005 (2005).
103. K. Ueda *et al.*, Appl. Phy. Lett. **86**, 172502 (2005).
104. H. Shimakage, K. Tsujimoto, Z. Wang, and M. Tonouchi, Appl. Phys. Lett. **86**, 072512 (2005).
105. H. Tanaka *et al.*, cond-mat/0407299 (unpublished).
106. H. Nakano *et al.*, cond-mat/0406622 (unpublished).
107. O. Astafiev *et al.*, Phys. Rev. B **69**, 180507(R) (2004).
108. P. Bertet *et al.*, Phys. Rev. B **70**, 100501(R) (2004).
109. A. Wallraff *et al.*, Phys. Rev. Lett. **95**, 060501 (2005).
110. I. Siddiqi *et al.*, Phys. Rev. Lett. **94**, 027005 (2005).
111. A. Lupascu *et al.*, Phys. Rev. Lett. **93**, 177006 (2004).
112. T. Yamamoto *et al.*, Nature **425**, 941 (2003).
113. R. McDermott *et al.*, Science **307**, 1299 (2005).
114. I. Chiorescu *et al.*, Nature **431**, 159 (2004).
115. A. Wallraff *et al.*, Nature **431**, 162 (2004).
116. D. I. Schuster *et al.*, Phys. Rev. Lett. **94**, 123602 (2005).
117. J. Johansson *et al.*, cond-mat/0510457 (unpublished).
118. K. Semba *et al.*, lT1271, poster presentation PB-F-86 in the 24th International Conference on Low Temperature Physics (LT24), Orlando, Florida, USA, (2005). (unpublished).
119. E. Hecht, *Optics* (Addison-Wesley, Reading, Mass., 1987).
120. A. Peres, *Quantum Theory: Concept and Methods* (Kluwer, Dordrecht, 1993).
121. D. Giulini *et al.*, *Decoherence and the Appearance of a Classical World in Quantum Theory* (Springer, Heidelberg, 1996).
122. W. Unruh, Phys. Rev. D **40**, 1053 (1989).
123. C. Cohen-Tannoudji, J. Dupont-Roc, and G. Grynberg, *Atom-Photon Interactions* (Wiley Interscience, New York, 1998).
124. E. Schrödinger, Naturwissenschaften **23**, 807, 823, 844 (1935).
125. V. Emery and A. Luther, Phys. Rev. B **9**, 215 (1974).
126. P. Anderson, Phys. Rev. Lett. **18**, 1049 (67).
127. R. J. Schoelkopf *et al.*, in *Quantum Noise*, *Nato ASI* (Kluwer, Dordrecht, 2002), Chap. Qubits as spectrometers of quantum noise, edited by Yu.V. Nazarov and Ya.M. Blanter.
128. R. Aguado and L. Kouwenhoven, Phys. Rev. Lett. **84**, 1986 (2000).
129. A. Cottet, Ph.D. thesis, Universite Paris 6, 2002.
130. J. Martinis *et al.*, Phys. Rev. B **67**, 094510 (2003).
131. C. Slichter, *Principles of magnetic resonance*, No. 1 in *Springer Series in Solid-State Sciences* (Springer, Berlin, 1996).

132. M. Grifoni, E. Paladino, and U. Weiss, Eur. Phys. J B **10**, 719 (1999).
133. L. Tian, S. Lloyd, and T. Orlando, Phys. Rev. B **65**, 144516 (2002).
134. A. Garg, J. Onuchic, and V. Ambegaokar, J. Chem. Phys. **83**, 4491 (1985).
135. A. Abragam, *Principles of nuclear magnetism*, Vol. 32 of *International series of monographs on physics* (Clarendon Press, Oxford, 1983).
136. S. Adler, Studies in history and philosophy of modern physics **34**, 135 (2003).
137. S. Nakaijma, Progr. Theor. Phys. **20**, 948 (1958).
138. R. Zwanzig, Phys. Rev. **124**, 983 (1961).
139. P. Argyres and P. Kelley, Phys. Rev. **134**, A98 (1964).
140. U. Weiss, *Quantum Dissipative Systems*, No. 10 in *Series in modern condensed matter physics*, 2 ed. (World Scientific, Singapore, 1999).
141. H. Spohn, Rev. Mod. Phys. **52**, 569 (1980).
142. G. Lindblad, Commun. Math. Phys. **48**, 119 (1976).
143. R. Alicki and K. Lendi, *Quantum dynamical semigroups and applications*, No. 286 in *Lecture notes in physics* (Springer, Berlin, 1976).
144. H. Charmichael, *An open systems approach to quantum optics* (Springer, Berlin, 1993).
145. C. Gardiner and P. Zoller, *Quantum Noise. A Handbook of Markovian and Non-Markovian Quantum Stochastic Methods with Applications to Quantum Optics, Springer series in synergetics* (Springer, Berlin, 1999).
146. D. Loss and D. DiVincenzo, cond-mat/0304118 (unpublished).
147. F. Bloch, Phys. Rev. **105**, 1206 (1957).
148. A. Redfield, IBM J Res. Develop. **1**, 19 (1957).
149. M. Plenio and P. Knight, Rev. Mod. Phys. **70**, 101 (1998).
150. J. Dalibard, Y. Castin, and K. Molmer, Phys. Rev. Lett. **68**, 580 (1992).
151. A. Korotkov, Phys. Rev. B **67**, 235408 (2003).
152. T. Costi, Phys. Rev. Lett. **80**, 1038 (1998).
153. S. Kehrein and A. Mielke, J. Stat. Phys. **90**, 889 (1998).
154. M. Keil and H. Schoeller, Phys. Rev. B **63**, 180302 (2001).
155. A. Leggett, Phys. Rev. B **30**, 1208 (1984).
156. C. van der Wal, F. Wilhelm, C. Harmans, and J. Mooij, Eur. Phys. J. B **31**, 111 (2003).
157. T. Robertson *et al.*, Phys. Rev. B **72**, 024513 (2005).
158. F. K. Wilhelm, Phys. Rev. B **68**, 060503(R) (2003).
159. S. K. and. S. Kehrein and J. von Delft, Phys. Rev. B **70**, 014516 (2004).
160. F. Wilhelm, S. Kleff, and J. van Delft, Chem. Phys. **296**, 345 (2004).
161. M. Thorwart, E. Paladino, and M. Grifoni, Chem. Phys. **296**, 333 (2004).
162. M. Goorden, M. Thorwart, and M. Grifoni, Phys. Rev. Lett. **93**, 267005 (2004).
163. A. Shnirman and G. Schoen, Phys. Rev. Lett. **57**, 15400 (1998).
164. C. van der Wal *et al.*, Science **290**, 773 (2000).
165. P. Joyez *et al.*, Journal of superconductivity **12**, 757 (1999).
166. W. Coffey, Y. Kalmykov, and J. Waldron, *The Langevin equation: With applications in physics chemistry and electrrical engineering*, No. 11 in *Series in contemporary chemical physics* (World Scientific, Singapore, 1996).

167. N. Prokovev and P. Stamp, Rep. Prog. Phys. **63**, 669 (2000).
168. H. Gassmann, F. Marquardt, and C. Bruder, Phys. Rev. E **66**, 041111 (02).
169. E. Paladino, L. Faoro, G. Falci, and R. Fazio, Phys. Rev. Lett. **88**, 228304 (2002).
170. K. Rabenstein, V. Sverdlov, and D. Averin, JETP Lett. **79**, 646 (2004).
171. Y. Galperin, B. Altshuler, and D. Shantsev, in *Fundamental problems of mesoscopic physics: Interaction and decoherence*, *NATO-ASI* (Plenum, New York, 2003), Chap. Low-frequency noise as a source of dephasing a qubit.
172. A. Shnirman, Y. Makhlin, and G. Schön, Phys. Scr. **T102**, 147 (2002).
173. P. Dutta and P. Horn, Rev. Mod. Phys. **53**, 497 (1981).
174. R. Wakai and D. van Harlingen, Phys. Rev. Lett. **58**, 1687 (1987).
175. N. Zimmermann, J. Cobb, and A. Clark, Phys. Rev. B **56**, 7675 (1997).
176. M. Weissman, Rev. Mod. Phys. **60**, 537 (1988).
177. N. van Kampen, *Stochastic processes in physics and chemistry* (Elsevier, Amsterdam, 1997).
178. L. Arnold, *Stochastische Differentialgleichungen* (Oldenbourg, Munich, 1973).
179. L. Viola and S. Lloyd, Phys. Rev. A **58**, 2733 (1998).
180. L. Viola, E. Knill, and S. Lloyd, Phys. Rev. Lett. **82**, 2417 (1999).
181. L. Viola, S. Lloyd, and E. Knill, Phys. Rev. Lett. **83**, 4888 (1999).
182. H. Carr and E. Purcell, Phys. Rev. **94**, 630 (1954).
183. H. Gutmann, F. Wilhelm, W. Kaminsky, and S. Lloyd, Quant. Inf. Proc. **3**, 247 (2004).
184. G. Weiss, *Aspects and Applications of the Random Walk* (North-Holland, Amsterdam, 1994).
185. H. Gutmann, W. Kaminsky, S. Lloyd, and F. Wilhelm, Phys. Rev. A **71**, 020302(R) (2005).
186. L. Faoro and L. Viola, Phys. Rev. Lett. **92**, 117905 (2004).
187. G. Falci, A. D'Arrigo, A. Mastellone, and E. Paladino, Phys. Rev. A **70**, R40101 (2004).
188. J. Majer *et al.*, Phys. Rev. Lett. **94**, 090501 (2005).
189. F. Strauch *et al.*, Phys. Rev. Lett. **91**, 167005 (2003).
190. D. Averin and C. Bruder, Phys. Rev. Lett. **91**, 057003 (2003).
191. J. Clarke *et al.*, Phys. Scr. **T102**, 173 (2002).
192. C. Hilbert and J. Clarke, J. Low Temp. Phys. **61**, 237 (1985).
193. A. Barenco, D. Deutsch, A. Ekert, and R. Josza, Phys. Rev. Lett. **74**, 4083 (1995).
194. J. Zhang, J. Vala, S. Sastry, and K. Whaley, Phys. Rev. A **67**, 042313 (2003).
195. Y. Makhlin, Quant. Inf. Proc. **1**, 243 (2002).
196. J. You, Y. Nakamura, and F. Nori, Phys. Rev. B **71**, 0242532 (2005).
197. M. Storcz and F. Wilhelm, Appl. Phys. Lett. **83**, 2389 (2003).
198. L. Levitov, T. Orlando, J. Majer, and J. Mooij, cond-mat/0108266 (unpublished).
199. P. Johnson *et al.*, Phys. Rev. B **67**, 020509 (2002).
200. A. Berkley *et al.*, Science **300**, 1548 (2003).
201. H. Xu *et al.*, Phys. Rev. Lett **94**, 027003 (2005).

202. J. You, J. Tsai, and F. Nori, Phys. Rev. Lett. **89**, 197902 (2002).
203. L. Wei, Y. x. Liu, and F. Nori, Europhys. Lett. **67**, 1004 (2004).
204. L. Wei, Y. x. Liu, and F. Nori, Phys. Rev. B **71**, 134506 (2005).
205. J. Lantz, M. Wallquist, V. Shumeiko, and G. Wendin, Phys. Rev. B **70**, 140507(R) (2004).
206. J. Siewert, R. Fazio, G. Palma, and E. Sciacca, J. Low Temp. Phys. **118**, 795 (2000).
207. Y. x. Liu, L. Wei, and F. Nori, Phys. Rev. A **71**, 063820 (2005).
208. M. Paternostro, G. Falci, M. Kim, and G. Palma, Phys. Rev. B **69**, 214502 (2004).
209. A. Spoerl *et al.*, quant-ph/0504202 (unpublished).
210. C. Rigetti, A. Blais, and M. Devoret, Phys. Rev. Lett. **94**, 240502 (2005).
211. A. Fowler, S. Devitt, and L. Hollenberg, Quant. Inf. Comp. **4**, 237 (2004).
212. I. Siddiqi *et al.*, cond-mat/0507548 (unpublished).
213. H. Tanaka, Y. Sekine, S. Saito, and H. Takayanagi, Physica C **368**, 300 (2002).

Controlling Three Atomic Qubits

H. Häffner[1,2], M. Riebe[1], F. Schmidt–Kaler[1], W. Hänsel[1], C. Roos[1], M. Chwalla[1], J. Benhelm[1], T. Körber[1], G. Lancaster[1], C. Becher[1], D.F.V. James[3] and R. Blatt[1,2]

[1] *Institut für Experimentalphysik, Innsbruck, Austria*

[2] *Institut für Quantenoptik und Quanteninformation, Österreichische Akademie der Wissenschaften, Austria*

[3] *Theoretical Division T–4, Los Alamos National Laboratory, Los Alamos NM 87545, United States*

We present a series of experiments where up to three ions held in a Paul trap are entangled, a given number of ions is selectively read out while conditional single–quantum–bit (qubit) operations are performed coherently on the remaining ion(s). Using these techniques, we demonstrate also a state transfer of a quantum bit from one ion to another one using two measurements and entanglement between an auxiliary ion and the target ion — also known as teleportation.

1. Introduction

Quantum information processing rests on the ability to control a quantum register[1]. In particular this includes initialization, manipulation and read–out of a set of qubits. After initialization, a sequence of quantum gate operations implements the algorithm, which usually generates multi–partite entangled states of the quantum register. Finally, the outcome of the computation is obtained by measuring the state of the individual quantum bits. In addition, for some important algorithms such as quantum error correction [1–5] and teleportation[6] a subset of the quantum register is read out selectively and subsequently operations on other qubits are carried out conditioned on the measurement result*.

*Indeed, such an error–correction scheme has been carried out recently in an ion trap by Chiaverini and co–workers[7] in an ion trap.

This idea of the selective read–out of a quantum register gains some additional appeal when carried out on an entangled register, because there the measurement process can be demonstrated in an extraordinary clear way. Producing entangled states is also the key ingredient for quantum information processing, and last but not least, such experiments realize some of the Gedanken experiments which helped significantly to develop quantum mechanics. Creation of entanglement with two or more qubits has already been demonstrated in the references [8–14]. However, so far only trapped ions have allowed to create entanglement in a completely deterministic way[9]. Our experiment allows the deterministic generation of 3–qubit entangled states and the selective read–out of an individual qubit followed by local quantum operations conditioned on the read–out. As we will see later, the selective read–out of the quantum register illuminates the measurement process in a very clear way.

The paper is organized as follows: first the deterministic creation of maximally entangled three–qubit states, specifically the Greenberger–Horne–Zeilinger (GHZ) state and the W–state, with a trapped–ion quantum computer is discussed[15]. In sections 4 and 5 we show how the qubits can be read out selectively and how GHZ– and W–states are affected by such local measurements. Next we demonstrate in section 6 operations conditioned on the read–out. This enables us to transform tripartite entanglement deterministically into bipartite entanglement with local operations and measurements. It also realizes a quantum eraser along the lines proposed in Ref.[16]. Finally, we implement a full deterministic quantum teleportation on demand[17] (see section 7).

2. Experimental setup

All experiments are performed in an elementary ion–trap quantum processor[18,19]. In order to investigate tripartite entanglement[20–22], we trap three $^{40}Ca^+$ ions in a linear Paul trap where they arrange themselves in a linear chain with an inter–ion distance of 5 μm. Qubits are encoded in a superposition of the $S_{1/2}$ ground state and the metastable $D_{5/2}$ state (lifetime $\tau \simeq 1.16$ s). Each ion–qubit is individually manipulated by a series of laser pulses on the $S \equiv S_{1/2}$ (m_j=-1/2) to $D \equiv D_{5/2}$ (m_j=-1/2) quadrupole transition near 729 nm employing narrowband laser radiation tightly focused onto individual ions in the string. The entire quantum register is prepared by Doppler cooling, followed by sideband ground state cooling of the center–of–mass vibrational mode ($\omega = 2\pi$ 1.2 MHz) as required for the controlled interaction of the ions according to the original proposal by Cirac and Zoller[23]. Finally, the ions' electronic qubit states are initialised in the S–state by optical pumping.

The operations which modify individual qubits and simultaneously connect a qubit to the bus (the center–of–mass mode) are performed by applying laser pulses on the carrier † or the "blue" sideband ‡ of the S→D transition. Qubit rotations can be written as unitary operations in the following way (c.f. [24]): carrier rotations are given by

$$R^C(\theta,\varphi) = \exp\left[i\frac{\theta}{2}\left(e^{i\varphi}\sigma^+ + e^{-i\varphi}\sigma^-\right)\right], \tag{1}$$

whereas transitions on the blue sideband are denoted as

$$R^+(\theta,\varphi) = \exp\left[i\frac{\theta}{2}\left(e^{i\varphi}\sigma^+a^\dagger + e^{-i\varphi}\sigma^-a\right)\right]. \tag{2}$$

Here $\sigma^\pm$ are the atomic raising and lowering operators which act on the electronic quantum state of an ion by inducing transitions from the $|S\rangle$ to $|D\rangle$ state and vice versa (notation: $\sigma^+ = |D\rangle\langle S|$). The operators a and $a^\dagger$ denote the annihilation and creation of a phonon at the trap frequency ω, i.e. they act on the motional quantum state. The parameter θ depends on the strength and the duration of the applied pulse and φ the relative phase between the optical field and the atomic polarization.

Ions are numbered in analogy to binary numbers such that the first ion is the right–most with the least significance. Defining the D–Level as logical 0, we obtain the following ordering of the basis: $|DDD\rangle, |DDS\rangle, |DSD\rangle \ldots$.

3. Preparing GHZ– and W–states

Entanglement of three qubits can be divided into two distinct classes[21]: GHZ–states and W–states. Choosing one representative of each class ($|GHZ\rangle \equiv (|SSS\rangle + |DDD\rangle)/\sqrt{2}$ and $|W\rangle \equiv (|DDS\rangle + |DSD\rangle + |SDD\rangle)/\sqrt{3}$) any pure entangled three qubit state can be created by single qubit operations on either $|GHZ\rangle$ or $|W\rangle$. We synthesize GHZ–states using a sequence of 10 laser pulses (Table 1) and W–states with a sequence of five laser pulses (see Table 2). These pulse sequences generate three–ion entangled states within less than 1 ms.

Full information on the three–ion entangled states is obtained by state tomography. For this the entangled states are subjected to 27 different sets

† $|S,n\rangle \rightarrow |D,n\rangle$ transition, i.e. no change of vibrational quantum number n, laser on resonance

‡ $|S,n\rangle \rightarrow |D,n+1\rangle$, laser detuned by $+\omega$

Table 1. Pulse sequence to create the GHZ–state ($|GHZ\rangle \equiv (|DDD\rangle + i|SSS\rangle)/\sqrt{2}$. For the definitions of $R_i^{C,+}(\theta, \varphi)$ see Eq. 1 and 2. First we apply a so–called beamsplitter pulse, creating a correlation between ion #1 and the bus mode (the phonon qubit). Ion #2 is flipped conditional on the phonon qubit with a CNOT–operation consisting of a phase gate[39] enclosed in two Hadamard–like operations. Finally the phonon qubit is mapped onto ion #3.

ion #1 (beamsplitter)	$R_1^+(\pi/2, 0)$	$R_1^C(\pi, \pi/2)$		
ion #2 (Hadamard)	$R_2^C(\pi/2, 0)$			
ion #2 (Phase gate)	$R_2^+(\pi, \pi/2)$	$R_2^+(\pi/\sqrt{2}, 0)$	$R^+(\pi, \pi/2)$	$R^+(\pi/\sqrt{2}, 0)$
ion #2 (Hadamard)	$R_2^C(\pi/2, \pi)$			
ion #3 (map)	$R_3^C(\pi, 0)$	$R_3^+(\pi, 0)$		

Table 2. Pulse sequence to create the W–state. First we apply an asymmetric beamsplitter pulse on ion #2 exciting the phonon mode with a probability of two–thirds. If the the phonon mode is excited, the second beamsplitter sequence removes the phonon with a probability of 0.5 and maps it onto ion #3. Finally, the last pulse maps the remaining phonon population onto ion #1 and we obtain $(|DDS\rangle + |DSD\rangle + |SDD\rangle)/\sqrt{3}$.

ion #2 (beamsplitter)	$R^+(2\arccos(1/\sqrt{3}), 0)$	
ion #3 (beamsplitter)	$R^C(\pi, \pi)$	$R^+(\pi/2, \pi)$
ion #1 (map)	$R^C(\pi, 0)$	$R^+(\pi, \pi)$

of single qubit operations before the read–out employing a CCD–camera. From this data all 64 entries of the density matrix are extracted with the methods described in [25,27]. In total 5000 experiments —corresponding to 200 s of measurement time— are sufficient to achieve an uncertainty of less than 2% for all density–matrix elements.

In Figs. 1 and 2 we show the experimental results for the density matrix elements of the GHZ– and W–states, $\rho_{|\mathrm{GHZ}\rangle}$ and $\rho_{|\mathrm{W}\rangle}$. The off–diagonal elements are observed with nearly equal height as the corresponding diagonal elements and with the correct phases. Fidelities of 76% for $\rho_{|\mathrm{GHZ}\rangle}$ and 83% for $\rho_{|\mathrm{W}\rangle}$ are obtained. The fidelity is defined as $|\langle\Psi_{\mathrm{ideal}}|\rho_{|\mathrm{exp}\rangle}|\Psi_{\mathrm{ideal}}\rangle|^2$ with Ψ_{ideal} denoting the ideal quantum state and $\rho_{|\mathrm{exp}\rangle}$ the experimentally determined density matrix.

All sources of imperfections have been investigated independently[19] and the measured fidelities are consistent with the known error budget. Note that for the W–state, coherence times greater than 200 ms were measured (exceeding the synthesis time by almost three orders of magnitude), while for the GHZ–state only ~1 ms was found. This is due to the W–states being a superposition of three states with the same energy. Thus, the W–states are not sensitive to the overall energy scale of the system and laser

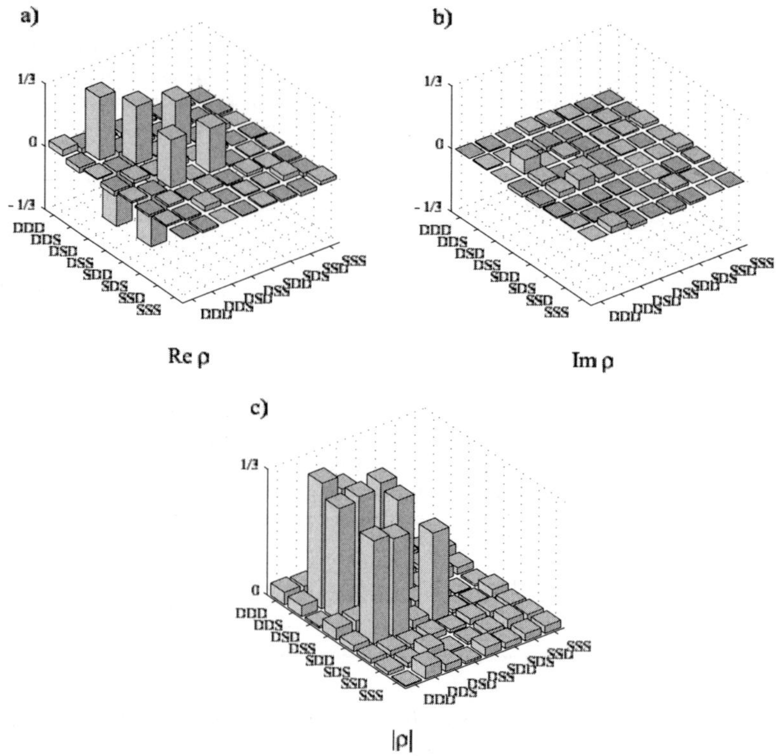

Fig. 1. Real part (a), imaginary part (b) and absolute values (c) of the density–matrix elements of the experimentally obtained W–state. The off–diagonal elements are of equal height as the diagonal elements and indicate the coherence between the different logical eigenstates {D, S}. The fidelity is calculated to be 83 %.

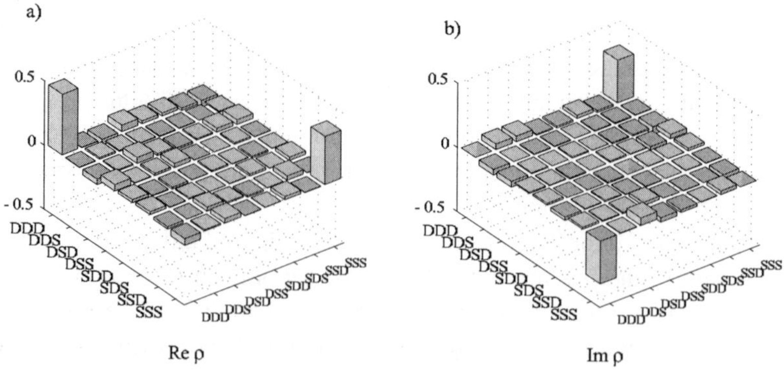

Fig. 2. Real (a) and imaginary (b) elements of a GHZ–states density matrix. The off–diagonal elements for SSS and DDD indicate the coherence. The fidelity was calculated to be 76 %.

frequency noise does not lead to dephasing. This is in strong contrast to the GHZ–state in Fig. 2 which is maximally sensitive to such perturbations. Similar behaviour has been observed previously with Bell-states[26,27].

4. Projection of the quantum states by measurement

Having tripartite entangled states available as a resource, we make use of individual ion addressing to project one of the three ions' quantum state to an energy eigenstate while preserving the coherence of the other two. Qubits are protected from being measured by transferring their quantum information into superpositions of levels which are not affected by the detection, that is a light scattering process on the $S_{1/2} \rightarrow P_{1/2}$–transition. In Ca^+, an additional Zeeman level D'≡ $D_{5/2}$ (m_j=-5/2) can be employed for this purpose. Thus, after the state synthesis, we apply two π pulses on the S→D' transition of ion #1 and #2, moving any S population of these ions into their respective D' level. The D and D' levels do not couple to the detection light at 397 nm (Fig. 3).

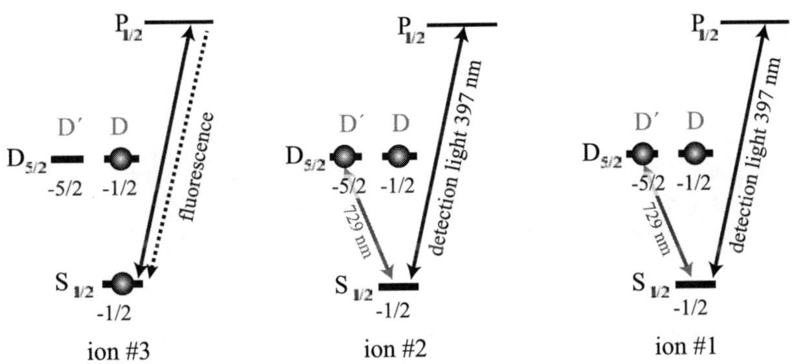

Fig. 3. Partial level scheme of the three Ca–ions. Only ion #3 is read out. Ion #1 and #2's quantum information is protected in the Zeeman manifold of the $D_{5/2}$–level, namely the $m_J = -1/2$ and $m_J = -5/2$ levels. Note that we have labelled the ions in analogy to a binary number representation from right to left.

Therefore, ion #3 can be read out using electron shelving[19]. After the selective readout a second set of π–pulses on the D' to S transition transfers the quantum information back into the original computational subspace {D, S}.

For a demonstration of this method, GHZ– and W–states are generated and the qubits #1 and #2 are mapped onto the {D, D'} subspace. Then, the state of ion #3 is projected onto S or D by scattering photons for a few microseconds on the S–P transition. In a first series of experiments, we did not distinguish whether ion #3 was projected into S or D. After remapping qubits #1 and #2 to the original subspace {S, D}, the tomography procedure is applied to obtain the full density matrix of the resulting three–ion state. As shown in Fig. 4c, the GHZ–state is completely destroyed, i.e.

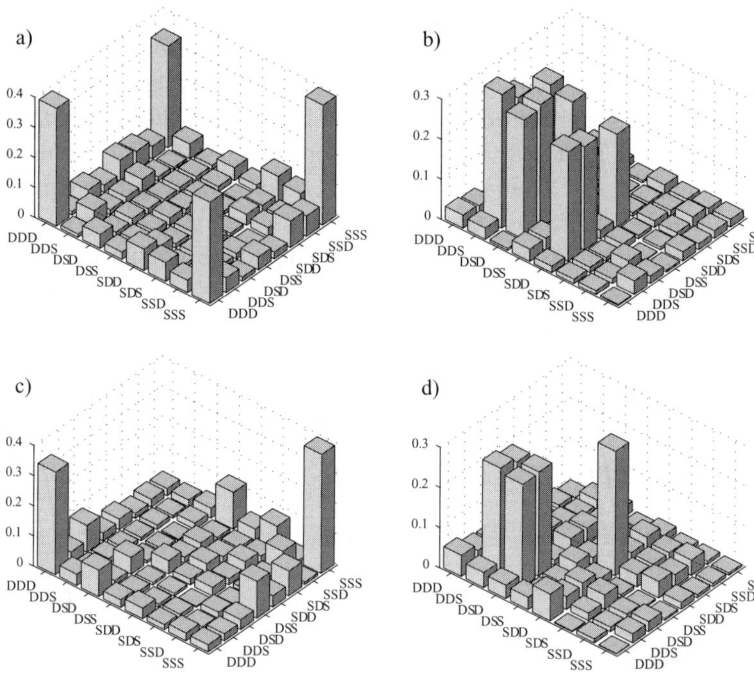

Fig. 4. Absolute values of density–matrix elements after measuring ion #3. (a) shows those of a GHZ–state before measuring and (c) after ion #3 is measured. The same for a W–state ((b) and (d)).

it is projected into a mixture of $|SSS\rangle$ and $|DDD\rangle$. In contrast, for the W–state, the quantum register remains partially entangled as coherences between ion #1 and #2 persist (Fig. 4c). Note that related experiments have been carried out with mixed states in NMR[14] and with photons[12].

5. Selective read–out of a quantum register

In a second series of experiments with W–states, we deliberately determine the third ion's quantum state prior to tomography: The ion string is now illuminated for 500 μs with light at 397 nm and its fluorescence is collected with a photomultiplier tube (Fig. 5a). Then, the state of ion #3 is known

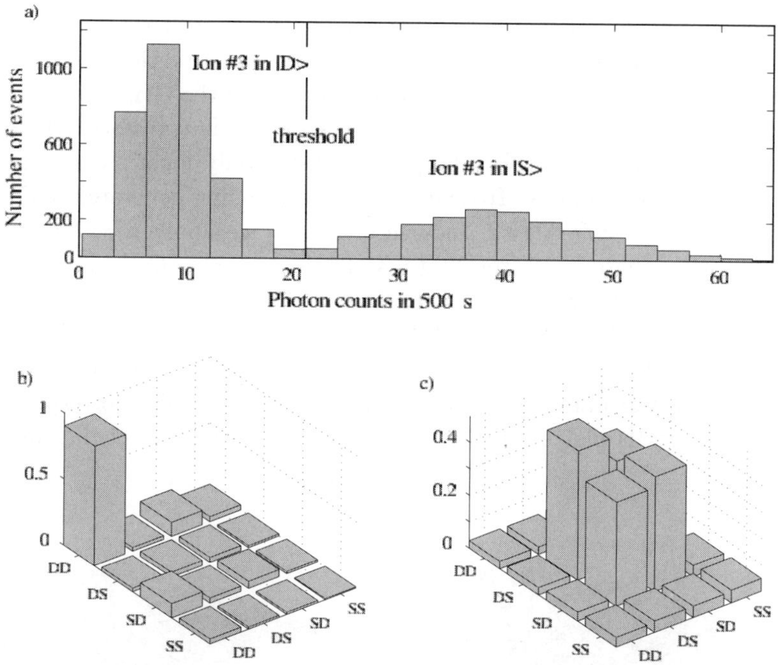

Fig. 5. (a) Histogram of photon counts within 500 s for ion #3 and threshold setting. (b) and (c) Density matrix of ion #1 and #2 conditioned upon the previously determined quantum state of ion #3. The absolute values of the reduced density–matrix elements are plotted for ion #3 measured in the S state (b) and ion #3 measured in the D state (c). Off–diagonal elements in (b) show the remaining coherences.

and subsequently we apply the tomographic procedure to ion #1 and #2 after remapping them to their {S, D} subspace. Depending on the state of ion #3, we observe the two density matrices presented in Fig. 5b and 5c. Whenever ion #3 was measured in D, ion #1 and #2 were found in a Bell state $(|SD\rangle + |DS\rangle)/\sqrt{2}$, with a fidelity of 82%. If qubit #3 was observed in S, the resulting state is $|DD\rangle$ with fidelity of 90%. This is a characteristic signature of $\mathrm{W} \equiv (|DDS\rangle + |DSD\rangle + |SDD\rangle)/\sqrt{3}$: In 1/3 of the cases, the measurement projects qubit #3 into the S state, and consequently the other two qubits are projected into D. With a probability of 2/3 however, the measurement shows qubit #3 in D, and the remaining quantum register is found in a Bell state[21]. Experimentally, we observe ion #3 in D in 65±2%

of the cases.

6. Conditioned single qubit operations

In section 4 we found that measuring a single qubit destroys the quantum nature of a GHZ–state completely. However, if prior to this the qubit is rotated into a different basis, the quantum nature of the GHZ–state can be partially preserved. Moreover, we can deterministically transform tripartite entanglement into bipartite entanglement using only local measurements and one–qubit operations. To demonstrate this, we first generate the GHZ–state $(|DSD\rangle + |SDS\rangle)/\sqrt{2}$. In a second step, we apply a $\pi/2$ pulse to ion #3, with phase $3\pi/2$, rotating a state $|S\rangle$ to $(|S\rangle - |D\rangle)/\sqrt{2}$ and $|D\rangle$ to $(|S\rangle + |D\rangle)/\sqrt{2}$, respectively. The resulting state of the three ions is $|D\rangle(|SD\rangle - |DS\rangle) + |S\rangle(|SD\rangle + |DS\rangle)/2$. A measurement of the third ion, resulting in $|D\rangle$ or $|S\rangle$, projects qubits #1 and #2 onto the state $(|SD\rangle - |DS\rangle)/\sqrt{2}$ or the state $(|SD\rangle + |DS\rangle)/\sqrt{2}$, respectively. The corresponding density matrix is plotted in Fig. 6a. With the information of the state of

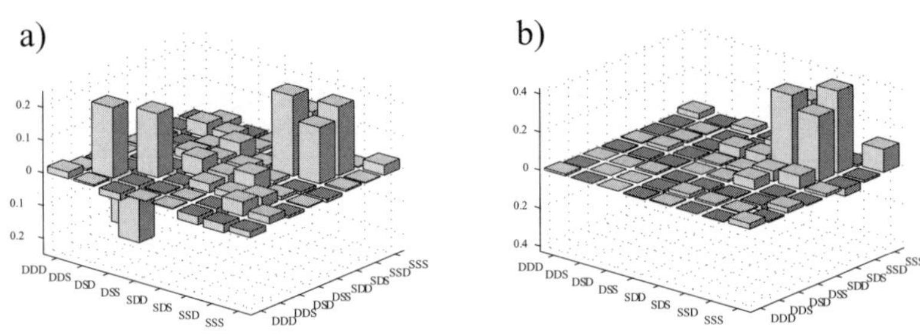

Fig. 6. (a) Real part of the density–matrix elements of the system after ion #1 of the GHZ–state $(|DSD\rangle + |SDS\rangle)/\sqrt{2}$ has been measured in a rotated basis. (b) Transformation of the GHZ–state $(|DSD\rangle + |SDS\rangle)/\sqrt{2}$ into the bipartite entangled state $|S\rangle(|DS\rangle + |SD\rangle)/\sqrt{2}$ by conditional local operations. Note the different vertical scaling of (a) and (b).

ion #3 available, we can now transform this mixed state into the pure state $|S\rangle(|SD\rangle + |DS\rangle)/\sqrt{2}$ by local operations only. Provided that ion #3 is found in $|D\rangle$, we perform a so–called Z–gate $(R^C(\pi, \pi/2)R^C(\pi, 0))$ on ion #2 to obtain $|D\rangle(|SD\rangle + |DS\rangle)/\sqrt{2}$. In addition, we flip the state of ion #3 to reset it to $|S\rangle$. Figure 6b shows that the bipartite entangled state $|S\rangle(|SD\rangle + |DS\rangle)/\sqrt{2}$ is produced with fidelity of 75%. This procedure can also be regarded as an implementation of a three–spin quantum eraser as proposed in[16].

Our results show that the selective read–out of a qubit in the quantum register indeed leaves all other qubits in the register untouched. In particular that means that for certain states entanglement can be preserved in the remaining part of the quantum register. In addition, even after such a measurement has taken place, single qubit rotations can be performed with high fidelity. Such techniques mark a first step towards the one–way quantum computer[28]. The implementation of unitary transformations conditioned on measurement results has great impact as it provides a way to implement active quantum–error–correction algorithms. In addition, we will show in the next sections that it allows for the realization of deterministic quantum teleportation.

7. Teleportation

Quantum teleportation exploits some of the most fascinating features of quantum mechanics, in particular *entanglement*, shedding new light on the essence of quantum information. It is possible to transfer the quantum information contained in a two–level system —a qubit— by communicating two classical bits and using entanglement. Thus quantum information can be broken down into a purely classical part and a quantum part.

Furthermore, teleportation is not merely a simple swapping of quantum states: it does not need a quantum channel to be open at the time the transfer is carried out. Instead it uses the non–local properties of quantum mechanics (entanglement), established by a quantum channel *prior to the generation of the state to be teleported.* Once that link has been established, an unknown state can be transferred at any later time using classical communication only. This is quite surprising since the quantum part of the transfer seems to have happenend before the state to be transferred exists. In addition to the motivation to demonstrate and to understand quantum physics, teleportation might also have considerable impact on a future quantum computer as it facilitates the scalability of many quantum computer designs[29].

Teleportation was already demonstrated with photonic qubits[30,31,11,32,33]. However, these experiments did not include complete two–photon Bell state measurements. In addition, successful teleportation events were established by selecting the data after completion of the experiment, searching for the subset of experiments in which the outcome of the measurement and a preset reconstruction operation were matched: i.e. teleportation was performed post–selectively. In contrast to this the experiment by Furusawa et *al.*[34] demonstrated unconditional teleportation of continuous variables. Similarly Nielsen et *al.*[35] implemented a deterministic teleportation algorithm with highly mixed states in a liquid–state NMR set–up.

Recently two groups realized quantum teleportation of atomic qubits. The Boulder group[36] teleported the quantum information contained in one Beryllium–ion to another one, while the Innsbruck group[17] used Calcium ions for the same purpose. Both groups trap their ions in Paul trap, however, pursue different approaches: in Boulder the qubits are encoded in the hyperfine structure of the ions, while in Innsbruck the qubit states are stored in superpositions of a ground and metastable electronic state. Furthermore the Boulder group uses segmented traps to perform the required selective read–out of the quantum register, whereas in Innsbruck tightly focused laser beams together with selective excitation of the Zeeman levels are employed for this purpose. Finally the Boulder group chose to work with a geometric phase gate[37], while the Innsbruck group uses composite pulses to realize the phase gate[18]. Despite these different approaches both experiments yield similar results. This demonstrates that ion traps are versatile devices for coherent state manipulation and quantum information processing.

The teleportation of a state from a source qubit to a target qubit requires three qubits: the sender's source qubit and an ancillary qubit that is maximally entangled with the receiver's target qubit providing the strong quantum correlation.

The quantum teleportation circuit is displayed in Fig. 7. The circuit is

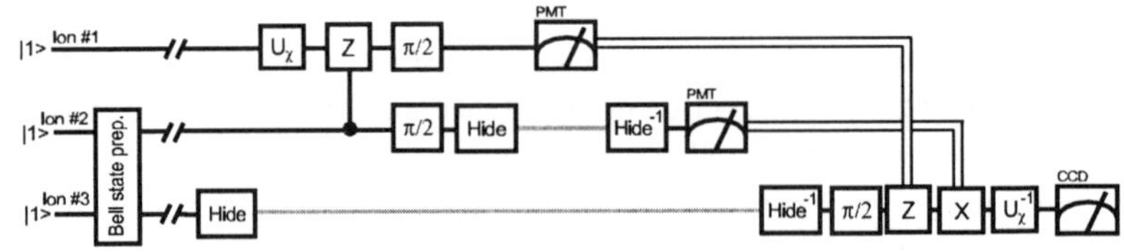

Fig. 7. The teleportation algorithm's quantum circuit. Double lines represent flow of classical information, whereas single lines flow of quantum information. The shaded lines indicate when a qubit is protected from detection light via so–called hiding–pulses. First ions #2 and #3 are entangled, creating the quantum link between the source region (ions #1 and #2) and the target ion (ion #3). Then after some waiting time the state to be teleported (on ion #1) is prepared via the unitary operation U_χ. A controlled Z–gate together with detection via a photomultiplier tube (PMT) realizes the Bell state measurement. After the reconstruction pulses the success of the teleportation is tested by appling the inverse preparation procedure before measuring the target ion on an intensified CCD–camera (Charge Coupled Device).

formally equivalent to the one proposed by Bennett et *al.* [6], but adapted to the ion–based quantum processor. It can be broken up into the following tasks:

(1) **Creation of Bell states**
A pulse sequence of three laser pulses (cf. Table 3) produces the Bell–state $(|DS\rangle + |SD\rangle/\sqrt{2}$. Tomography[38,27] of this state shows a fidelity of up to 96% for the entangling operation. Similary to the W–states above this Bell state constists of a superpositions of states with the same energy. Indeed, we observe that the lifetime of this Bell state approaches the fundamental limit given by the spontaneous decay rate of the metastable $D_{5/2}$–level of 1.2 s[27]. Now, after the quantum link between the source and the target region is established, we prepare a test state χ via a single qubit operation $U\chi$ on the source ion.

(2) **Rotation into the Bell–basis**
The Bell–state measurement is accomplished by rotating the basis of the source and the ancilla ions into the Bell basis before the actual read–out of the qubits. This rotation is implemented with a controlled–Z (phase) gate and appropriate single qubit operations. The experimental implementation of the controlled–Z–gate is described in ref. [18]. To illustrate the rotation into the Bell–basis more easily, we will use in the following a zero–controlled–not (0–CNOT) gate as a substitute for the controlled Z–gate: suppose one has the Bell state $(|DS\rangle + |SD\rangle)/\sqrt{2}$ (note that we use the convention $|D\rangle \equiv |0\rangle$ and $|S\rangle \equiv |1\rangle$), then application of a 0–CNOT followed by a $\pi/2$–Carrier–Pulse on the control bit (the leftmost bit) yields:

$$(|DS\rangle + |SD\rangle)/\sqrt{2} \xrightarrow{0-CNOT} (|DD\rangle + |SD\rangle)/\sqrt{2} \tag{3}$$

$$= (|D\rangle + |S\rangle)|D\rangle/\sqrt{2} \xrightarrow{R_C^{\mathrm{C}}(\pi/2,0)} |SD\rangle \tag{4}$$

The pulse $R_C^{\mathrm{C}}(\pi/2, 0)$ denotes a single qubit rotation of the control bit. Now we have mapped the Bell state $|DS\rangle + |SD\rangle$ to $|SD\rangle$. Similarly all other Bell states are mapped onto orthogonal logical eigenstates. Therefore a measurement in the logical eigenbasis yields now a precise knowledge of the original Bell state.

(3) **Selective read–out of the quantum register and conditional quantum gates**
The measurement process must preserve the coherence of the target qubit, ion #3. Thus, the state of ion #3 is hidden by transferring it to a superposition of levels which are not affected by the detection

light. We employ an additional Zeeman level of the $D_{5/2}$ manifold for this purpose. Applying now laser light at 397 nm for 250 μs to the ion crystal, only the ion in question can fluoresce, and that only if it is the $S_{1/2}$–state[15]. This hiding technique is also used to sequentially read out ion #1 and ion #2 with a photomultiplier tube (see Fig. 3). Instead of using a CCD–camera (which can easily distinguish between different ions), we prefer to take advantage of the fast electronic read–out capabilities of a photo–multiplier tube. This ensures a reaction on the measurement result within the single qubit coherence time. A digital electronic circuit counts the number of detected photons and compares it to the threshold (less than 6 detected photons indicate that the ion is in the $D_{5/2}$ level).

Conditioned on the measurement result, we apply single qubit rotations on the target ion[15]. This is implemented by using a classical AND–gate between the output of the electronic circuit which has stored the measurement result and the output of a digital board on which the reconstruction pulses are programmed. Thus, we apply the appropriate unitary qubit rotation, $-i\sigma_y$, $-i\sigma_z$, $i\sigma_x$, or 1 (with Pauli operators σ_k) to reconstruct the state in the target ion #3, obtaining χ on ion #3. Note that $-i\sigma_z$ is realized by applying σ_y followed by σ_x. This has the advantage that we can apply σ_x if ion #1 is measured to be in $|D\rangle$ and σ_z if ion #2 is measured to be in $|D\rangle$. Thus the logic of the controll electronic remains quite simple.

The whole pulse sequence is displayed in Table 3. In contrast to Fig.7, here also spin echo pulses are included. The conditioned pulses #31,32,33 are applied only if less than 6 photon detection events were recorded during the respective detection time of 250 μs. The phase ϕ for the pulses is fixed during all experiments. It is used to compensate for the 50 Hz related magnetic field fluctuations during the execution of the teleportation algorithm.

To obtain directly the fidelity of the teleportation, we perform on ion #3 the operation $U^{-1}\chi$, which is the inverse of the unitary operation used to create the input state $|\chi\rangle$ from state $|S\rangle$ (see pulses #9 and #34 in Table 3). The teleportation is successful if and only if ion #3 is always found in $|S\rangle$. The teleportation fidelity, given by the overlap $F = \langle S|U^{-1}\chi\rho_{\text{exp}}U\chi|S\rangle$, is plotted in Fig. 8 for all four test states $\{|S\rangle, |D\rangle, |S\rangle + |D\rangle, |S\rangle + i|D\rangle\}$. The obtained fidelities range from 73% to 76%. Teleportation based on a completely classical resource instead of a quantum entangled resource yields a maximal possible fidelity of 66.7% [41] (dashed line in Fig. 8). Note that

Table 3. To implement the teleportation, we use pulses on carrier transitions $R_i^C(\theta,\varphi)$ and $R_i^H(\theta,\varphi)$ (no change of the motional state of the ion crystal) and, additionally, on the blue sideband $R_i^+(\theta,\varphi)$ (change of the motional state) on ion i. The index C denotes carrier transitions between the two logical eigenstates, while the index H labels transitions from the $S-$ to the D'–level. For the definitions of $R_i^{C,H,+}(\theta,\varphi)$ see Eqs. 1 and 2.

		Action	Comment
	1	Light at 397 nm	Doppler preparation
	2	Light at 729 nm	Sideband cooling
	3	Light at 397 nm	Optical pumping
Entangle	4	$R_3^+(\pi/2, 3\pi/2)$	Entangle ion #3 with motional qubit
	5	$R_2^C(\pi, 3\pi/2)$	Prepare ion #2 for entanglement
	6	$R_2^+(\pi, \pi/2)$	Entangle ion #2 with ion #3
	7	Wait for 1μs – 10 000 μs	Stand–by for teleportation
	8	$R_3^H(\pi, 0)$	Hide target ion
	9	$R_1^C(\vartheta\chi, \varphi\chi)$	Prepare source ion #1 in state χ
Rotate into Bell–basis	10	$R_2^+(\pi, 3\pi/2)$	Get motional qubit from ion #2
	11	$R_1^+(\pi/\sqrt{2}, \pi/2)$	Composite pulse for phasegate
	12	$R_1^+(\pi, 0)$	Composite pulse for phasegate
	13	$R_1^+(\pi/\sqrt{2}, \pi/2)$	Composite pulse for phasegate
	14	$R_1^+(\pi, 0)$	Composite pulse for phasegate
	15	$R_1^C(\pi, \pi/2)$	Spin echo on ion #1
	16	$R_3^H(\pi, \pi)$	Unhide ion #3 for spin echo
	17	$R_3^C(\pi, \pi/2)$	Spin echo on ion #3
	18	$R_3^H(\pi, 0)$	Hide ion #3 again
	19	$R_2^+(\pi, \pi/2)$	Write motional qubit back to ion #2
	20	$R_1^C(\pi/2, 3\pi/2)$	Part of rotation into Bell–basis
	21	$R_2^C(\pi/2, \pi/2)$	Finalize rotation into Bell basis
Read–out	22	$R_2^H(\pi, 0)$	Hide ion #2
	23	PMDetection for 250 μs	Read out ion #1 with photomultiplier
	24	$R_1^H(\pi, 0)$	Hide ion #1
	25	$R_2^H(\pi, \pi)$	Unhide ion #2
	26	PMDetection for 250 μs	Read out ion #2 with photomultiplier
	27	$R_2^H(\pi, 0)$	Hide ion #2
	28	Wait 300 μs	Let system rephase; part of spin echo
	29	$R_3^H(\pi, \pi)$	Unhide ion #3
	30	$R_3^C(\pi/2, 3\pi/2 + \phi)$	Change basis
Recon-struction	31	$R_3^C(\pi, \phi)$	$i\sigma_x$ } $= -i\sigma_z$ conditioned on
	32	$R_3^C(\pi, \pi/2 + \phi)$	$-i\sigma_y$ } PMDetection #1
	33	$R_3^C(\pi, \phi)$	$i\sigma_x$ conditioned on PMDetection #2
	34	$R_3^C(\vartheta\chi, \varphi\chi + \pi + \phi)$	Inverse of preparation of χ with offset ϕ
	35	Light at 397 nm	Read out ion #3 with camera

this classical boundary holds only if no assumptions on the states to be teleported are made. If one restricts oneself to only the four test states, strategies exist which use no entanglement and yield fidelities of 78% [40].

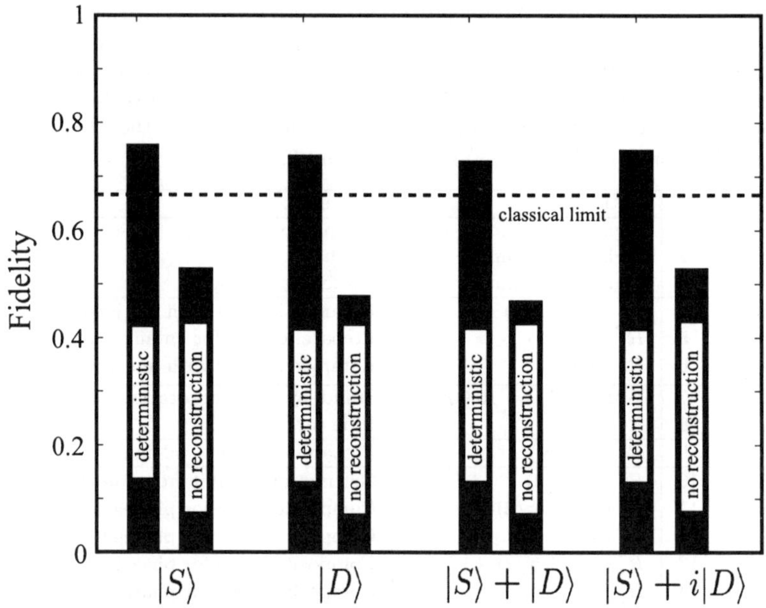

Fig. 8. Result of the teleportation: The four test states are teleported with fidelities of 76%, 74%, 73%, and 75%, respectively (bars labelled 'reconstruction'). For each input state 300 single teleportation experiments were performed. The error of each entry, estimated from quantum projection noise, is 2.5%. The bars labelled 'no reconstruction' show the results if the reconstruction operations are omitted, yielding an average fidelity of 49.6%. The optimum teleportation obtainable by purely classical means and no assumptions about the initial states reaches a fidelity of 66.7% (dashed line).

However, each of these strategies must be designed for a specific test state ensemble to work properly. Note also that, in order to rule out out hidden variable theories, a fidelity in excess of 0.87 is required[42].

For comparison, we also show data where the reconstruction pulses were not applied. Without the classical information obtained from the Bell state measurement, the receiver's state is maximally mixed, i.e. there is no information available on the source state. Also, the measurement outcome of ions #1 and #2 does not contain any information about the initial state. Indeed we find each possible result with an equal probability of 0.25 ± 0.036, independent of the test input states. Note that only with both the receiver's qubit and the result of the Bell measurement, the initial state can be retrieved.

We emphasize that the conditional, deterministic reconstruction step, in combination with the complete Bell state analysis, is one of the crucial improvements with respect to former experimental realizations of quantum teleportation. Furthermore, after the teleportation procedure the state χ is always available and may be used for further experiments.

To emphasize the role of the shared entangled pair as a resource, we store the Bell state for some time and then use it only later (after up to 20 ms) for teleportation. For waiting times of up to 20 ms (exceeding the time we require for the teleportation by a factor of 10) we observe no decrease in the fidelity. For longer waiting times, we expect the measured heating of the ion crystal of smaller than 1 phonon/100 ms to reduce the fidelity significantly. This is because for a successful rotation into the Bell–basis we require the phonon number in center–of–mass mode of the ion string to be in the ground state.

The obtainable fidelity is limited mainly by dephasing mechanisms. The most obvious one is frequency fluctuations of the laser driving the qubit transition, and magnetic field fluctuations. Since these fluctuations are slow compared to the execution time of 2 ms, they can be cancelled to some degree with spin echo techniques[43]. However, during the algorithm we have to use different pairs of states to encode the quantum information, one of which being only sensitive to magnetic field fluctuations while the other one being sensitive to both laser and magnetic field fluctuations. To overcome these complications, two spin echo pulses are introduced (see Table 3). Their optimal position in time was determined with numerical simulations. From measurements we estimate that the remaining high frequency noise reduces the fidelity by about 5%. Another source of fidelity loss is an imperfect AC–Stark shift compensation. AC–Stark compensation is needed to get rid of the phase shifts introduced by the laser driving the weak sideband transition due to the presence of the strong carrier transitions[44]. Recent measurements suggest that an imperfect compensation as introduced by the incorrect determination of the sideband frequency by only 100 Hz lead to a loss of teleportation fidelity on the order of 5%.

Imperfect state detection as introduced by a sub–optimal choice for the threshold (6 instead of 3 counts) was analyzed later to contribute on the order of 3% to the fidelity loss. However, the biggest contribution to the read–out error stems from an incorrect setting of the hiding pulse frequency and strength. It reduced the fidelity by 7%.

Addressing errors on the order of 3–4% were estimated via numerical simulations to reduce the fidelity by about 6%. The addressing errors were measured by comparing the Rabi flopping frequency between neighboring ions and correspond to a ratio of 10^{-3} in intensity between the addressed ion and the other ones.

Treating these estimated error sources independently (multiplying the success probabilities) yields an expected fidelity of 77% in good agreement with the experimental findings.

In conclusion, we described an experiment demonstrating teleportation of atomic states. The experimental procedures might be applied in future quantum information processing networks: with long lived entangled states as a resource, quantum teleportation can be used for the distribution of quantum information between different nodes of the network.

We gratefully acknowledge support by the European Commission (QUEST and QGATES networks), by the ARO, by the Austrian Science Fund (FWF). H.H. acknowledges funding by the Marie–Curie–program of the European Union. T.K. acknowledges funding by the Lise–Meitner program of the FWF. We are also grateful for discussions with A. Steinberg and H. Briegel.

References

1. M. A. Nielsen, I. L. Chuang, Quantum Computation and Quantum Information (Cambridge Univ. Press, Cambridge, 2000).
2. P. W. Shor, Phys. Rev. A **52**, R2493 (1995).
3. A. M. Steane, Phys. Rev. Lett. **77**, 793 (1996).
4. C. H. Bennett, D. P. DiVincenzo, J.A. Smolin, W. K. Wootters, Phys. Rev. A **54**, 3824 (1996).
5. A. M. Steane, Nature **399**, 124 (1999).
6. Bennett, C. H., G. Brassard, C. Crépeau, R. Jozsa, A. Peres, and W. K. Wootters, Phys. Rev. Lett. **70**, 1895 (1993).
7. J. Chiaverini, D. Leibfried, T. Schaetz, M.D. Barrett, R.B. Blakestad, J. Britton, W.M. Itano, J.D. Jost, E. Knill, C. Langer, R. Ozeri & D.J. Wineland, Nature **432**, 602 (2004).
8. A. Rauschenbeutel, G. Nogues; S. Osnaghi, P. Bertet, M. Brune, J.–M. Raimond, S. Haroche, Science **288**, 2024 (2000).
9. C.A. Sackett, D. Kielpinski, B.E. King, C. Langer, V. Meyer, C.J. Myatt, M. Rowe, Q.A. Turchette, W.M. Itano, D.J. Wineland, and C. Monroe, Nature **404**, 256 (2000).
10. J. W. Pan, D. Bouwmeester, M.Daniell, H. Weinfurter, A. Zeilinger, Nature **403**, 515 (2000).
11. J. W. Pan, M. Daniell, S. Gasparoni, G. Weihs, A. Zeilinger, Phys. Rev. Lett. **86**, 4435 (2001).
12. M. Eibl, N. Kiesel, M. Bourennane, C. Kurtsiefer, H. Weinfurter, Phys. Rev. Lett. **92**, 077901 (2004).
13. Z. Zhao, Y.–A. Chen, A.–N. Zhang, T. Yang, H. Briegel, J.–W. Pan, Nature **430**, 54 (2004).

14. G. Teklemariam, E. M. Fortunato, M. A. Pravia, Y. Sharf, T. F. Havel, D. G. Cory, A. Bhattaharyya, and J. Hou , Phys. Rev. A **66**, 012309 (2002).
15. C. F. Roos, M. Riebe, H. Häffner, W. Hänsel, J. Benhelm, G. P. T. Lancaster, C. Becher, F. Schmidt–Kaler, R. Blatt, Science **304**, 1478 (2004).
16. R. Garisto and L. Hardy, Phys. Rev. A 60, 827, (1999).
17. M. Riebe, H. Häffner, C. F. Roos, W. Hänsel, J. Benhelm, G. P. T. Lancaster, T. W. Körber, Nature **429**, 734 (2004).
18. F. Schmidt–Kaler, H. Häffner, M. Riebe, S. Gulde, G. P. T. Lancaster, T. Deuschle, C. Becher, C. F. Roos, J. Eschner & R. Blatt, Nature **422**, 408 (2003).
19. F. Schmidt-Kaler, H. Häffner, S. Gulde, M. Riebe, G. P.T. Lancaster, T. Deuschle, C. Becher, W. Hänsel, J. Eschner, C. F. Roos, R. Blatt, Appl. Phys. B: Lasers and Optics **77**, 789 (2003).
20. D. M. Greenberger, M. Horne, A. Zeilinger, in Bellś Theorem, Quantum Theory, and Conceptions of the Universe, edited by M. Kafatos (Kluwer Academic, Dordrecht, 1989).
21. W. Dür, G. Vidal, J. I. Cirac, Phys. Rev. A **62**, 062314 (2000).
22. A. Zeilinger, M. A. Horne D. M. Greenberger, NASA Conf. Publ. No. 3135 National Aeronautics and Space Administration, Code NTT, Washington, DC, (1997).
23. J.I. Cirac and P. Zoller, Phys. Rev. Lett. **74**, 4091 (1995)
24. M. Šašura and V. Bužek, J. Mod. Opt. **49**, 1593 (2002).
25. D. F. V. James, P. G. Kwiat, W. J. Munro, A. G. White, Phys. Rev. A **64**, 052312 (2001).
26. D. Kielpinski, V. Meyer, M.A. Rowe, C.A. Sackett, W.M. Itano, C. Monroe, D.J. Wineland, Science **291**, 1013 (2001).
27. C. F. Roos, G. P. T. Lancaster, M. Riebe, H. Häffner, W. Hänsel, S. Gulde, C. Becher, J. Eschner, F. Schmidt–Kaler, R. Blatt, Phys. Rev. Lett. **92**, 220402 (2004).
28. R. Raussendorf and H. J. Briegel, Phys. Rev. Lett. 86, 5188 (2001).
29. D. Gottesman D. & I.L. Chuang, Nature **402**, 390 (1999).
30. D. Bouwmeester, J–W. Pan, K. Mattle, M. Eible, H. Weinfurter, and A. Zeilinger , Nature **390**, 575 (1997).
31. D. Boschi, S. Branca, F. DeMartini, L. Hardy and S. Popescu , Phys. Rev. Lett. **80**, 1121 (1998).
32. I. Marcikic, H. de Riedmatten, W. Tittel, H. Zbinden, & N. Gisin, Nature **421**, 509–513 (2003).
33. D. Fattal, E. Diamanti, K. Inoue and Y. Yamamoto, Phys. Rev. Lett. **92**, 037904 (2004).
34. A. Furusawa, J.L. Sørensen, S.L. Braunstein, C.A. Fuchs, H.J. Kimble, E.S. Polzik, Science **282**, 706 (1998).
35. M. A. Nielsen, E. Knill & R. Laflamme, Nature **396**, 52 (1998).
36. M. D. Barrett, J. Chiaverini, T. Schaetz, J. Britton, W.M. Itano, J.D. Jost, E. Knill, C. Langer, D. Leibfried, R. Ozeri & D.J. Wineland, Nature **429**, 737 (2004).
37. D. Leibfried , B. DeMarco, V. Meyer, D. Lucas, M. Barrett, J. Britton, W.

M. Itano, B. Jelenković, C. Langer, T. Rosenband & D. J. Wineland, Nature **422**, 412 (2003).

38. K. Vogel and H. Risken, Phys. Rev. A **40**, 2847 (1989).
39. S. Gulde, M. Riebe, G.P.T. Lancaster, C. Becher, J. Eschner, H. Häffner, F. Schmidt–Kaler, I.L. Chuang & R. Blatt, Nature **421**, 48 (2003).
40. Steven van Enk, private communication.
41. S. Massar and S. Popescu, Phys. Rev. Lett. **74**, 1259 (1995).
42. N. Gisin, Physics Letters A **210**, 157 (1996).
43. E.L. Hahn, Phys. Rev. **77**, 746 (1950)
44. H. Häffner, S. Gulde, M. Riebe, G. Lancaster, C. Becher, J. Eschner, F. Schmidt–Kaler, and R. Blatt, Phys. Rev. Lett. **90**, 143602–1–4 (2003).

Liquid-State NMR Quantum Computer: Hamiltonian Formalism and Experiments

Yasushi KONDO* and Mikio NAKAHARA

Department of Physics, Kinki University
Higashi-Osaka 577-8502, Japan
**E-mail: kondo@phys.kindai.ac.jp*

Shogo TANIMURA

Graduate School of Engineering, Osaka City University
Osaka 558-8585, Japan

A quantum algorithm is represented as a unitary operator acting on a qubit space and is realized as a time development operator of a controlled quantum system. In a liquid-state NMR quantum computer, spins of nuclei in isolated molecules dissolved in a solvent work as qubits and they are controlled by radio frequency pulses. We reexamine the Hamiltonian describing the spin dynamics and formulate the principle of a liquid-state NMR quantum computer from the viewpoint of Hamiltonian dynamics. We executed the Deutsch-Jozsa algorithm and the Grover database search algorithm as well as a pseudopure state preparation with two-qubit homonucleus molecules to demonstrate the validity of our formalism. Finally we discuss whether a liquid-state NMR quantum computer currently fulfills the DiVincenzo criteria or not.

1. Introduction

Quantum computation has been recently attracting a lot of attention since it is expected to solve some of computationally hard problems for a conventional digital computer.[1] Various physical systems for its realization have been proposed to date. Above all, a liquid-state NMR (nuclear magnetic resonance) is considered to be most successful owing to a beautiful demonstration of Shor's factorization algorithm.[2] Disadvantages of a liquid-state NMR quantum computer have been also pointed out.[3] For example, the number of qubits is possibly limited up to about ten because of poor spin polarization at room temperature and overlapping chemical shifts. In spite of these limitations, the liquid-state NMR quantum computer is the best test bench to study implementation of quantum algorithms at the moment.

Therefore it is important to understand its dynamics from a Hamiltonian viewpoint.

Suppose we would like to implement a quantum algorithm whose unitary matrix representation is $U_{\rm alg}$. Our task is to find the control parameters $\gamma(t)$ in the Hamiltonian H such that the time development operator

$$U[\gamma(t)] = \mathcal{T} \exp\left[-\frac{i}{\hbar}\int_0^T H(\gamma(t))dt\right] \tag{1}$$

is equal to $U_{\rm alg}$, where $\mathcal{T}$ stands for the time-ordered product.

This contribution is organized as follows. In Section 2 we explain the principle of NMR measurements. We put emphasis on the role of a Hamiltonian that governs spin dynamics. In Section 3 control of two-qubit molecules is discussed. Since two qubits are minimal for performing nontrivial quantum computation, two-qubit systems are worth discussing in detail. Issues related to the density matrix are discussed in Section 4. In Section 5 quantum computations, by employing cytosine as an example, are discussed. Section 6 is devoted to discussion on the *DiVincenzo criteria*[4] and summary. First four sections lay foundation of NMR quantum computing while Sections 5 and 6 deal with more advanced topics. Other aspects of liquid-state NMR quantum computing are available in excellent reviews.[5–8]

2. NMR Primer

Here we describe a typical experimental setup for NMR measurement and explain its theoretical basis. We exclusively analyze molecules with a single spin-$\frac{1}{2}$ nucleus in this section while molecules with more spins will be studied in the following sections.

2.1. *Experimental setup*

NMR is an experimental method to measure and control a state of nuclear spins in a molecule. Molecules in the sample tube are moving and rotating rapidly and randomly and hence the intermolecular spin-spin interactions are averaged out to vanish while intramolecular spin-spin interactions must be taken into account when writing down the Hamiltonian. Suppose a static magnetic field is applied on a sample that contains nuclear spins. Then the eigenstates of the spins are defined with respect to the magnetic field. If the spins are not in one of the eigenstates, they exhibit a precession, which is observed as a rotating magnetization of the sample. When an oscillating magnetic field with frequency equal to the energy difference in two spin

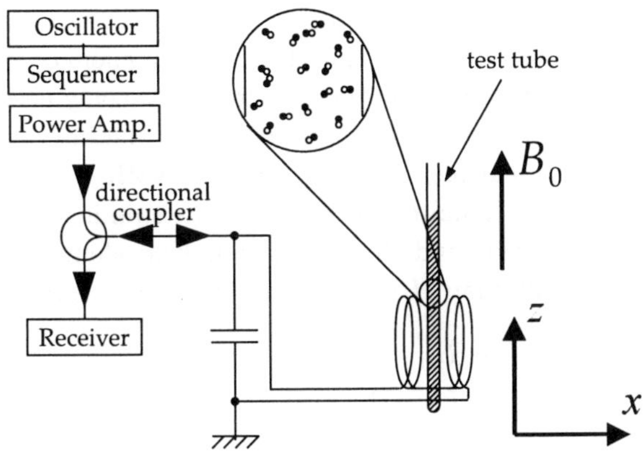

Fig. 1. NMR setup. Molecules in the test tube are moving and rotating rapidly and randomly.

states is applied in addition, the spin state continuously changes. The final state of spins is also measured as a rotating magnetization.

An experimental setup for NMR measurement, shown schematically in Fig. 1, consists of three parts. The first part contains a superconducting magnet system to provide a homogeneous constant magnetic field and a field gradient coil system to produce temporally controlled field gradients. The second is a resonance circuit to apply rf (radio frequency) magnetic fields on the sample and pick up rf signals from rotating magnetization. The third is an assembly of electronic circuits to produce rf pulses and detect rf signals.[9,10]

The electric oscillator generates a continuous alternating current with the radio frequency $\omega_{\rm rf}$ and the sequencer modulates the oscillating current to generate rf pulses. A typical temporal duration of a pulse, which is called a pulse width, is of the order of 10 μs. The rf pulses are amplified and fed into the resonance circuit, which generates rf magnetic fields on the sample in the test tube. The directional coupler is inserted to prevent direct transmission of the rf pulses into the receiver. The rotating magnetization of the sample induces signals at the coil, which are lead to the receiver and detected.

2.2. *Hamiltonian for a single spin*

There are a large number, typically $\sim 10^{20}$, of molecules in the test tube. Sample molecules interact with each other and also with the solvent molecules. These interactions cause random motion of the molecules. However, these motions are very rapid, of the order of 1 ps, and thus spin-spin interactions among these molecules are temporally averaged to vanish. In other words, we may assume that spin degrees of freedom of each molecule are effectively isolated from each other. Therefore, we only need to consider a single molecule Hamiltonian and totally ignore the interactions among molecules. This is a great merit of a liquid-state NMR.

Let us consider the Hamiltonian H of a single spin in a static magnetic field along the z-direction and the rf magnetic field along the x-direction, see Fig. 1. The Hamiltonian H is decomposed into two terms as

$$\begin{aligned} H &= H_0 + H_{\rm rf} \\ H_0 &= -\hbar\omega_0 I_z, \quad H_{\rm rf} = 2\hbar\omega_1 \cos(\omega_{\rm rf}t - \phi) I_x, \end{aligned} \tag{2}$$

where $I_i = \sigma_i/2$ and σ_i is the Pauli matrix indexed by $i = 1,2,3 = x,y,z$. The parameter ω_0 is proportional to the strength of the magnetic field B_0 and is given by $\omega_0 = \gamma B_0$ with the gyromagnetic ratio γ of the nucleus. The first term H_0 describes action of the static magnetic field on the spin. In the static magnetic field the spin precesses with the Larmor frequency ω_0. The second term $H_{\rm rf}$ describes the action of the rf pulses on the spin and hence it is controllable. Control parameters are the time-dependent field strength ω_1, the frequency $\omega_{\rm rf}$, and the phase ϕ of the rf field. Note that $H_{\rm rf}$ represents a linearly oscillating field along the x-axis, not a circularly rotating field.

The state of ensemble of molecules is described by a density matrix

$$\rho = \sum_{i=0}^{1}\sum_{j=0}^{1} \rho_{ij} |i\rangle \langle j|, \tag{3}$$

where $|0\rangle$ and $|1\rangle$ denote the eigenstate of the spin $\uparrow$ (lower energy state) and $\downarrow$ (higher energy state), respectively. The ensemble average of an observable M is given by

$$\mathrm{tr}(\rho M) = \sum_{i,j=0}^{1} \rho_{ij} \langle j| M |i\rangle . \tag{4}$$

The density matrix $\rho_{\rm th}$ of the thermal state without rf magnetic fields is (see, Eq. (2))

$$\begin{aligned} \rho_{\rm th} &= \frac{e^{-(H_0/k_BT)}}{\mathrm{tr}\left[e^{-(H_0/k_BT)}\right]} = \frac{e^{-(-\hbar\omega_0 I_z/k_BT)}}{\mathrm{tr}\left[e^{-(-\hbar\omega_0 I_z/k_BT)}\right]} \\ &\simeq \frac{1}{2}\left(I + \frac{\hbar\omega_0}{k_BT} I_z\right). \end{aligned} \tag{5}$$

The above approximation may be justified since $\hbar\omega_0/k_BT \sim 10^{-5}$ at room temperature.

The time development of $\rho(t)$ is determined by the Liouville equation

$$i\hbar\frac{d\rho}{dt} = [H, \rho] = H\rho - \rho H. \tag{6}$$

We introduce a time-dependent unitary operator $U(t)$ and define the unitary transformation of the density matrix as

$$\tilde{\rho} = U\rho U^\dagger. \tag{7}$$

Then Eq. (6) becomes

$$i\hbar\frac{d\tilde{\rho}}{dt} = [\tilde{H}, \tilde{\rho}] \tag{8}$$

with the transformed Hamiltonian

$$\tilde{H} = UHU^\dagger - i\hbar U\frac{dU^\dagger}{dt}. \tag{9}$$

If we take the unitary transformation

$$U = \exp(-i\omega_{\rm rot}I_z t), \tag{10}$$

then $\tilde{\rho}$ represents the spin state viewed in a frame rotating with the frequency $\omega_{\rm rot}$ around the z-axis. The transformed Hamiltonian becomes $\tilde{H} = \tilde{H}_0 + \tilde{H}_{\rm rf}$ with

$$\tilde{H}_0 = -\hbar(\omega_0 - \omega_{\rm rot})I_z$$

and

$$\begin{aligned}\tilde{H}_{\rm rf} &= \hbar\omega_1\left[\{\cos\phi + \cos(\phi - 2\omega_{\rm rot}t)\}I_x + \{\sin\phi - \sin(\phi - 2\omega_{\rm rot}t)\}I_y\right] \\ &\quad - \hbar\omega_1\left[\cos(\phi - \omega_{\rm rot}t) - \cos(\phi - \omega_{\rm rf}t)\right]\begin{pmatrix} 0 & e^{-i\omega_{\rm rot}t} \\ e^{i\omega_{\rm rot}t} & 0\end{pmatrix}.\end{aligned}$$

The above Hamiltonian is simplified if we take $\omega_{\rm rot} = \omega_{\rm rf}$ as

$$\begin{aligned}\tilde{H}_0 &= -\hbar(\omega_0 - \omega_{\rm rf})I_z \\ \tilde{H}_{\rm rf} &= \hbar\omega_1\left[\{\cos\phi + \cos(\phi - 2\omega_{\rm rf}t)\}I_x + \{\sin\phi - \sin(\phi - 2\omega_{\rm rf}t)\}I_y\right].\end{aligned}$$

The typical time scale $1/\omega_1$ of the spin dynamics is much longer than the period of the oscillating field such that $1/\omega_1 \gg 1/\omega_{\rm rf}$. Thus the terms rapidly oscillating with frequency $2\omega_{\rm rf}$ in $\tilde{H}_{\rm rf}$ are averaged to vanish. Therefore, $\tilde{H}_{\rm rf}$ is approximated as

$$\tilde{H}_{\rm rf} = \hbar\omega_1\left(\cos\phi I_x + \sin\phi I_y\right).$$

in the rotating frame. Moreover, when the resonance condition $\omega_0 = \omega_{\rm rf}$ is satisfied, $\tilde{H}_0$ vanishes and the total Hamiltonian reduces to

$$\tilde{H} = \tilde{H}_{\rm rf} = \hbar\omega_1 (\cos\phi I_x + \sin\phi I_y)\,. \tag{11}$$

Therefore, the spin dynamics becomes simpler if it is viewed in the rotating frame with $\omega_{\rm rot} = \omega_{\rm rf} = \omega_0$.

The fact that the factor 2 in the Hamiltonian $H_{\rm rf}$ (Eq. (2)) disappears in the transformed Hamiltonian $\tilde{H}_{\rm rf}$ (Eq. (11)) is physically understood as follows.[9] An oscillating rf magnetic field is considered to be a superposition of two rotating fields of $\pm\omega_{\rm rf}$ and the effect of the rotating field with $-\omega_{\rm rf}$ is averaged out to vanish in our rotating frame.

2.3. *Dynamics*

Let us apply the previous argument to analyze the spin dynamics. The density matrix $\rho_{\rm th}$ in Eq. (5) represents the thermal state in the laboratory frame. It is transformed to $\tilde{\rho}_{\rm th}$ in the rotating frame as

$$\begin{aligned}\tilde{\rho}_{\rm th} &= U\,\frac{1}{2}\left(I + \frac{\hbar\omega_0}{k_B T}\,I_z\right)U^\dagger \\ &= \exp(-i\omega_0 I_z t)\,\frac{1}{2}\left(I + \frac{\hbar\omega_0}{k_B T}\,I_z\right)\exp(i\omega_0 I_z t) \\ &= \frac{1}{2}\left(I + \frac{\hbar\omega_0}{k_B T}\,I_z\right).\end{aligned} \tag{12}$$

Actually $\tilde{\rho}_{\rm th}$ is identical to $\rho_{\rm th}$ since both $I/2$ and I_z commute with $\exp(\pm i\omega_0 t I_z)$. Suppose that an rf magnetic field with phase $\phi = 0$ is applied for the period of τ so that $\omega_1\tau = \pi/2$. The reduced Hamiltonian in the rotating frame is

$$\tilde{H} = \hbar\omega_1 I_x, \tag{13}$$

which represents an effective magnetic field parallel to the x-axis. The time development operator (1) now becomes

$$\exp\left(-\frac{i}{\hbar}\int_0 \tau\hbar\omega_1 I_x dt\right) = \exp\left(-i\frac{\pi}{2}I_x\right). \tag{14}$$

The density matrix $\tilde{\rho}_{\rm rf}$ obtained after the rf pulse is applied to $\tilde{\rho}_{\rm th} = \rho_{\rm th}$ then becomes

$$\begin{aligned}\tilde{\rho}_{\rm rf} &= \exp\left(-i\frac{\pi}{2}I_x\right)\frac{1}{2}\left(I + \frac{\hbar\omega_0}{k_B T}\,I_z\right)\exp\left(i\frac{\pi}{2}I_x\right) \\ &= \frac{1}{2}\left(I - \frac{\hbar\omega_0}{k_B T}\,I_y\right).\end{aligned} \tag{15}$$

Physically, the magnetization is turned from the z-direction to the $-y$-direction by the effective field parallel to the x-axis in the rotating frame. We call this pulse a $\pi/2$-pulse around the x-axis. Then, the density matrix $\rho_{\rm rf}$ in the laboratory frame after the application of this rf pulse is

$$\begin{aligned}\rho_{\rm rf} &= U^\dagger \frac{1}{2}\left(I - \frac{\hbar\omega_0}{k_B T} I_y\right) U \\ &= \exp(i\omega_0 I_z t)\,\frac{1}{2}\left(I - \frac{\hbar\omega_0}{k_B T} I_y\right)\exp(-i\omega_0 I_z t) \\ &= \frac{1}{2}I - \frac{\hbar\omega_0}{4k_B T}\begin{pmatrix} 0 & -ie^{i\omega_0 t} \\ ie^{-i\omega_0 t} & 0 \end{pmatrix}. \end{aligned} \tag{16}$$

We calculate the expectation value of magnetization in the x-direction, which generates the signal from the sample. The result is proportional to

$$\mathrm{tr}(\sigma_x \rho_{\rm rf}) = -\frac{\hbar\omega_0}{2k_B T}\sin(\omega_0 t). \tag{17}$$

This represents the magnetization oscillating with a frequency ω_0.

2.4. *Quantum state tomography*

The density matrix is Hermitian and thus generally parametrized as

$$\tilde{\rho} = \frac{1}{2}\begin{pmatrix} 1+c & a-ib \\ a+ib & 1-c \end{pmatrix} = \frac{I}{2} + aI_x + bI_y + cI_z \tag{18}$$

by real parameters $a, b, c \simeq \hbar\omega_0/2k_B T$. Note that $I/2$ is included for normalization ($\mathrm{tr}\tilde{\rho} = \mathrm{tr}(I/2) = 1$). By measuring all the parameters a, b, and c, we know the state of the system completely. The technique to measure them is called *quantum state tomography*. The density matrix ρ in the laboratory frame at time t is

$$\begin{aligned}\rho(t) &= U^\dagger(t)\tilde{\rho}\,U(t) \\ &= \exp(i\omega_0 t I_z)\tilde{\rho}\exp(-i\omega_0 t I_z) \\ &= \frac{1}{2}\begin{pmatrix} 1+c & (a-ib)e^{i\omega_0 t} \\ (a+ib)e^{-i\omega_0 t} & 1-c \end{pmatrix}. \end{aligned} \tag{19}$$

Therefore, the x-component of the magnetization in the laboratory frame is

$$M(t) \propto \mathrm{tr}(\sigma_x\,\rho) = a\cos\omega_0 t + b\sin\omega_0 t. \tag{20}$$

The density matrix after the application of a $\pi/2$-pulse along the x-axis in the rotating frame is

$$e^{-i\pi I_x/2}\tilde{\rho}\,e^{i\pi I_x/2} = \frac{I}{2} + aI_x - cI_y + bI_z \tag{21}$$

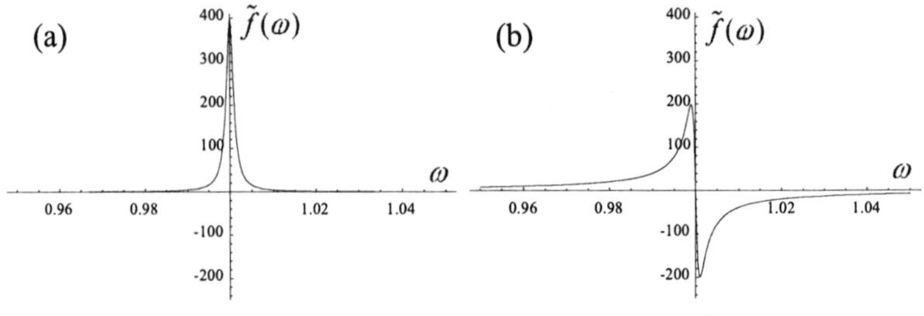

Fig. 2. Cosine Fourier transformations $\hat{f}(\omega)$ of free induction decay signals (a) $f(t) = e^{-t/T_2}\cos\omega_0 t$ and (b) $f(t) = e^{-t/T_2}\sin\omega_0 t$. Here $\omega_0 = 1$ and $1/T_2 = 0.001$.

and it follows from Eq. (20) that the x-component of the magnetization in the laboratory frame is

$$M(t) \propto a\cos\omega_0 t - c\sin\omega_0 t. \tag{22}$$

Similarly the density matrix after the application of a $\pi/2$-pulse along the y-axis in the rotating frame is

$$e^{-i\pi I_y/2}\tilde{\rho}\,e^{i\pi I_y/2} = \frac{I}{2} + cI_x + bI_y - aI_z \tag{23}$$

and the x-component of the magnetization in the laboratory frame is

$$M(t) \propto c\cos\omega_0 t + b\sin\omega_0 t. \tag{24}$$

Therefore, a, b, and c can be measured with a phase sensitive detector, as we show in the next section.

2.5. *Measurements*

There are relaxation mechanisms in reality and a signal induced in the coil lasts only for a limited period of time. The signal is called the free induction decay (FID) signal after pulses and its time constant is denoted as T_2. The signal is well approximated by

$$f(t) = e^{-t/T_2}\cos(\omega_0 t + \phi). \tag{25}$$

The cosine Fourier transformation of $f(t) = e^{-t/T_2}\cos\omega_0 t$ and $f(t) = e^{-t/T_2}\sin\omega_0 t$ are shown in Fig. 2.

From FID signals generated by the oscillating magnetization (20), we find the coefficients a and b by Fourier transformation. Similarly, by applying the pulses along the x- and y-directions, we measure the other coefficients in Eqs. (22) and (24). The components in the signal with $\cos\omega_0 t$ and $\sin\omega_0 t$ are often called *real* and *imaginary* parts, respectively, in NMR.

2.6. *Operations in rotating frame*

2.6.1. *Unitary operations*

Suppose that an rf magnetic field with the phase ϕ is applied to a spin during a period of τ such that $\omega_1\tau = \theta$. The Hamiltonian (11) in the rotating frame defines the time development operator as

$$\begin{aligned} U_\phi(\theta) &= \exp\left[-\frac{i}{\hbar}\int_0^\tau \hbar\omega_1(I_x\cos\phi + I_y\sin\phi)dt\right] \\ &= \exp\left[-i\theta(I_x\cos\phi + I_y\sin\phi)\right]. \end{aligned} \tag{26}$$

The thermal state evolves into

$$\begin{aligned} \tilde{\rho}_{\mathrm{rf}} &= U_\phi(\theta)\tilde{\rho}_{\mathrm{th}}U_\phi^\dagger(\theta) \\ &= \frac{I}{2} + \frac{\hbar\omega_0}{2k_BT}\left[\sin\theta(I_x\sin\phi - I_y\cos\phi) + I_z\cos\theta\right] \end{aligned} \tag{27}$$

after the pulse is applied. This shows that the magnetization has been rotated by an angle θ around the axis $(\cos\phi, \sin\phi, 0)$ in the rotating frame.

We sometimes need a transformation

$$U_z(\psi) = \exp(-i\psi I_z) \tag{28}$$

that rotates the spin by an angle ψ around the z-axis. The generator I_z does not exist in the Hamiltonian (11) in the rotating frame and hence it seems to be impossible to rotate a spin around the z-axis by any pulse. There are, however, several ways to rotate a spin around the z-axis.

One way is to apply a sequence of three pulses

$$U_0(\pi/2)U_{\pi/2}(\theta)U_\pi(\pi/2).$$

The product is

$$\exp\left(-i\frac{\pi}{2}I_x\right)\exp\left(-i\theta I_y\right)\exp\left(i\frac{\pi}{2}I_x\right) = \exp\left(-i\theta I_z\right) = U_z(\theta). \tag{29}$$

Another sequence

$$U_{\pi/2}(\pi/2)U_0(-\theta)U_{3\pi/2}(\pi/2) = U_z(\theta) \tag{30}$$

also rotates a spin by θ around the z-axis.

Another method to deal with the spin rotation around the z-axis in the rotating frame uses the relation

$$U_\phi(\theta)U_z(\psi) = U_z(\psi)U_{\phi-\psi}(\theta), \tag{31}$$

which is easily proved. Repeated use of the above relation makes it possible to shift $U_z(\psi)$ toward the end of the pulse sequence. The rotation around the z-axis at the end (or beginning) of the pulse sequence can be reinterpreted as a phase of a quantum system, which is not observable. Therefore the rotation around the z-axis may be, in fact, dropped.

Finally we have a complete set of one-qubit operations, which turn a spin to any directions.

2.6.2. *Operators in actual pulse sequences*

In practical applications, pulses that rotate a spin by $\pi/2$ around the x-, $-x$-, y-, $-y$-, z- and $-z$-axes are commonly used. Let us summarize here the properties of these operations. The symbols $X, Y, Z, \bar{X}, \bar{Y}$ and $\bar{Z}$ denote the rotations of the spin by $\pi/2$ around the x-, y-, z-, $-x$-, $-y$- and $-z$-axes, respectively. Their explicit forms are

$$\begin{aligned}
X &= U_0(\pi/2) = \tfrac{1}{\sqrt{2}}\begin{pmatrix} 1 & -i \\ -i & 1 \end{pmatrix}, \quad & \bar{X} &= U_\pi(\pi/2) = \tfrac{1}{\sqrt{2}}\begin{pmatrix} 1 & i \\ i & 1 \end{pmatrix}, \\
Y &= U_{\pi/2}(\pi/2) = \tfrac{1}{\sqrt{2}}\begin{pmatrix} 1 & -1 \\ 1 & 1 \end{pmatrix}, \quad & \bar{Y} &= U_{3\pi/2}(\pi/2) = \tfrac{1}{\sqrt{2}}\begin{pmatrix} 1 & 1 \\ -1 & 1 \end{pmatrix}, \\
Z &= U_z(\pi/2) = \tfrac{1}{\sqrt{2}}\begin{pmatrix} 1-i & 0 \\ 0 & 1+i \end{pmatrix}, \quad & \bar{Z} &= U_z(-\pi/2) = \tfrac{1}{\sqrt{2}}\begin{pmatrix} 1+i & 0 \\ 0 & 1-i \end{pmatrix}.
\end{aligned} \tag{32}$$

The following relations are useful to simplify pulse sequences:

$$\begin{array}{llll}
XY\bar{X} = Z, & \bar{Y}XY = Z, & \bar{X}\bar{Y}X = Z, & Y\bar{X}\bar{Y} = Z \\
\bar{X}YX = \bar{Z}, & YX\bar{Y} = \bar{Z}, & X\bar{Y}\bar{X} = \bar{Z}, & \bar{Y}\bar{X}Y = \bar{Z} \\
XY = ZX, & XY = YZ, & \bar{Y}X = XZ, & Y\bar{X} = ZY \\
XZ = Z\bar{Y}, & \bar{Y}Z = Z\bar{X}, & \bar{X}Z = ZY, & YZ = ZX \\
XZZ = ZZ\bar{X}, & & YZZ = ZZ\bar{Y}. &
\end{array} \tag{33}$$

The Hadamard gate H_1, which often appears in quantum algorithms, takes the form

$$H_1 = \frac{1}{\sqrt{2}}\begin{pmatrix} 1 & 1 \\ 1 & -1 \end{pmatrix} \tag{34}$$

and is constructed as a product

$$H = iYZZ. \tag{35}$$

Note that H_1 is often *approximated* by Y in actual NMR pulse sequences for quantum algorithms. This approximation breaks down if the input state is a superposition of $|0\rangle$ and $|1\rangle$.

2.6.3. *Field gradient*

A field gradient is the inhomogeneity in strength of the static magnetic field that is applied to the sample. Field gradients are employed in quantum computation to introduce non-unitary operations in preparation of a pseudopure state, for example.

We consider single spin dynamics in the presence of a field gradient. A field gradient modifies the system Hamiltonian (2) as

$$H_0 = -\hbar\omega_0(1 + \nabla z)I_z, \tag{36}$$

where ∇ is the normalized strength of the field gradient along the z-direction. We assume that the sample extends over $-L_0 < z < L_0$. In the frame rotating with frequency ω_0, the relevant Hamiltonian depends on z as

$$\tilde{H}_0 = -\hbar\nabla\omega_0 z I_z \tag{37}$$

and thus the time development operator (1) is

$$\begin{aligned} U_G(z) &= \exp\left(-\frac{i}{\hbar}\int_0^\tau -\hbar\nabla\omega_0 z I_z dt\right) \\ &= \exp\left(i\nabla\omega_0 z I_z \tau\right). \end{aligned} \tag{38}$$

After application of the field gradient for a time interval τ, the density matrix $\tilde{\rho}_G(z)$ in the rotating frame becomes z-dependent as

$$\begin{aligned} \tilde{\rho}_G(z) &= U_G(z)\tilde{\rho}U_G^\dagger(z) \\ &= \exp\left(i\nabla\omega_0 z\tau I_z\right)\left(\frac{I}{2} + aI_x + bI_y + cI_z\right)\exp\left(-i\nabla\omega_0 z\tau I_z\right) \\ &= \frac{I}{2} + cI_z + \begin{pmatrix} 0 & (a-ib)e^{i\nabla\omega_0 z\tau} \\ (a+ib)e^{-i\nabla\omega_0 z\tau} & 0 \end{pmatrix} \end{aligned} \tag{39}$$

for the initial density matrix $\tilde{\rho} = I/2 + aI_x + bI_y + cI_z$. Since the signal picked up by the coil is a sum of the contributions from all the sample molecules, $\tilde{\rho}_G(z)$ should be averaged over the sample size $-L_0 < z < L_0$. The third term in the last line of Eq. (39) is averaged to vanish if the condition $L_0\nabla\omega_0\tau \gg 1$ is satisfied. Therefore, application of a field gradient is equivalent to a non-unitary operation that removes off-diagonal elements of the density matrix as

$$D_G : \tilde{\rho} = \frac{I}{2} + aI_x + bI_y + cI_z \mapsto \frac{1}{2L_0}\int_{-L_0}^{L_0} U_G(z)\tilde{\rho}U_G^\dagger(z)dz = \frac{I}{2} + cI_z. \tag{40}$$

2.6.4. *Bloch-Siegert effect*

Bloch-Siegert effect was originally introduced to take into account the effect of a counter rotating rf field which is time-averaged to be zero in Eq. (11).[11] Ramsey then generalized it to the case when two or more rf fields with different frequencies are applied to the sample.[12] Although multiple rf fields are dealt in both the original and generalized Bloch-Siegert effects, its essential effect can be considered as an influence of a widely off-resonance rf field with the angular velocity $\omega_0+\Delta$ on a spin with the resonance frequency ω_0.

Let us start with a Hamiltonian in the laboratory frame,

$$H_{\rm BS} = -\hbar\omega_0 I_z + 2\hbar\omega_1 \cos\left((\omega_0+\Delta)t\right) I_x, \tag{41}$$

where we took $\omega_{\rm rf} = \omega_0 + \Delta$ and $\phi = 0$ in (11). To simplify our calculation, we take

$$U = \exp\left[-i(\omega_0+\Delta)tI_z\right].$$

Then, we obtain the Hamiltonian in the rotating frame with the angular velocity $\omega_0+\Delta$,

$$\tilde{H}_{\rm BS} = \hbar\Delta I_z + \hbar\omega_1 I_x, \tag{42}$$

where rapidly oscillating terms are averaged to vanish. Since (42) is time-independent, the time-development operator Eq. (1) is calculated as

$$\begin{aligned}
&\exp\left(-\frac{i}{\hbar}\int_0^\tau \tilde{H}_{\rm BS}dt\right) \qquad (43)\\
&= \begin{pmatrix} \cos\left(\frac{\tau\Delta}{2}\sqrt{1+\epsilon^2}\right) - i\frac{\sin\left(\frac{\tau\Delta}{2}\sqrt{1+\epsilon^2}\right)}{\sqrt{1+\epsilon^2}} & -i\frac{\epsilon\sin\left(\frac{\tau\Delta}{2}\sqrt{1+\epsilon^2}\right)}{\sqrt{1+\epsilon^2}} \\ -i\frac{\epsilon\sin\left(\frac{\tau\Delta}{2}\sqrt{1+\epsilon^2}\right)}{\sqrt{1+\epsilon^2}} & \cos\left(\frac{\tau\Delta}{2}\sqrt{1+\epsilon^2}\right) + i\frac{\sin\left(\frac{\tau\Delta}{2}\sqrt{1+\epsilon^2}\right)}{\sqrt{1+\epsilon^2}} \end{pmatrix} \\
&= e^{-i\tau\Delta\sqrt{1+\epsilon^2}I_z} - i\sin\left(\frac{\tau\Delta}{2}\sqrt{1+\epsilon^2}\right)\begin{pmatrix} -1+\frac{1}{\sqrt{1+\epsilon^2}} & \frac{\epsilon}{\sqrt{1+\epsilon^2}} \\ \frac{\epsilon}{\sqrt{1+\epsilon^2}} & 1-\frac{1}{\sqrt{1+\epsilon^2}} \end{pmatrix},
\end{aligned}$$

where $\epsilon = \omega_1/\Delta$. When we assume $\Delta \gg \omega_1$, i.e. $\epsilon \ll 1$, (43) becomes

$$\exp\left(-\frac{i}{\hbar}\int_0^\tau \tilde{H}_{\rm BS}dt\right) = e^{-i\tau\Delta\sqrt{1+\epsilon^2}I_z}. \tag{44}$$

The meaning of Eq. (44) is (a) the applied off-resonance rf field does not rotate the spin and (b) the Larmor frequency is apparently shifted by $\Delta(1-\sqrt{1+\epsilon^2})$. Although the shift of the Larmor frequency is small, $|\tau\Delta(1-\sqrt{1+\epsilon^2})|$ is not necessarily small because τ can be large.

In NMR quantum computation with homonucleus molecules, the Bloch-Siegert effect sometimes have to be taken into account. Even though a soft selective pulse is easily designed so that it does not practically rotate the other spins, it may still affect the phases of the other spins, see Eq. (44). Therefore, it causes the errors in the phase of the subsequent (or, simultaneously applied) pulses on the other spins.

Let us consider three examples of the Bloch-Siegert effect on a two-spin homonucleus molecule. We call the two spins A with the Larmor frequency $\omega_{0\mathrm{A}}$ and B with $\omega_{0\mathrm{B}} = \omega_{0\mathrm{A}} + \Delta$.

(1) Square π-pulse with the pulse width $2 \cdot 2\pi/\Delta$:
The pulse on B causes the approximate phase shift

$$-\frac{1}{2}\frac{(\pi/(2 \cdot 2\pi/\Delta))^2}{\Delta} \cdot (2 \cdot 2\pi/\Delta) = -\frac{\pi}{8} \simeq -0.39$$

on A.

(2) Square π-pulse with the pulse width $4 \cdot 2\pi/\Delta$:
The pulse on B causes the approximate phase shift

$$-\frac{1}{2}\frac{(\pi/(4 \cdot 2\pi/\Delta))^2}{\Delta} \cdot (4 \cdot 2\pi/\Delta) = -\frac{\pi}{16} \simeq -0.20$$

on A.

(3) Square $\pi/2$-pulse with the pulse width $4 \cdot 2\pi/\Delta$:
The pulse on B causes the approximate phase shift

$$-\frac{1}{2}\frac{((\pi/2)/(4 \cdot 2\pi/\Delta))^2}{\Delta} \cdot (4 \cdot 2\pi/\Delta) = -\frac{\pi}{64} \simeq -0.05$$

on A.

The soft pulse which we employ in the following sections is a Gaussian pulse[9] and its Bloch-Siegert effect is about 1.7 times larger than that of the square pulse with the same pulse width. We mainly employed the $\pi/2$-Gaussian pulses with the pulse width of $4 \cdot 2\pi/\Delta$, and thus we can, as a first approximation, neglect the Bloch-Siegert effect in the examples we will discuss in the following sections.

More sophisticated shaped pulses such as UBURP[9] are sometimes employed by others. We note that they affect the NMR quantum computation more than Gaussian pulses do since they have much larger peak strengths of the rf field than Gaussian pulses.

3. Hamiltonians for two-qubit molecules

We consider molecules with two spins here since any nontrivial quantum computation requires at least two qubits. There are two classes of molecules with multiple spins; heteronucleus molecules and homonucleus molecules. Heteronucleus molecules are easy to control. However the number of qubits therein is practically limited to two or three due to the limitation inherent in an NMR spectrometer.

3.1. *Heteronucleus molecules*

3.1.1. *Experimental Setup*

The NMR setup for heteronucleus molecules is shown schematically in Fig. 3. A typical example of a two-qubit heteronucleus molecule is ^{13}C-labeled chloroform, for which ^{13}C and H nuclei are qubits. Because the resonance frequencies $\omega_{0,i}$, $i = 1, 2$ for the first and second spins, respectively, are largely different, two resonance circuits and two assemblies of electronic circuits are required. The difference in the resonance frequencies, $\Delta\omega_0 = (\omega_{0,2} - \omega_{0,1})$, allows us to address individual spins in quantum computation. We assume, without loss of generality, that $\Delta\omega_0 > 0$.

The oscillator 1 (2) generates an oscillating electric current with frequency $\omega_{0,1}$ ($\omega_{0,2}$). The sequencer 1 (2) modulates the oscillating current to generate rf pulses. A typical temporal duration of a pulse, called a pulse width, is on the order of 10 μs. The rf pulses are amplified and fed into the resonance circuit 1 (2), which generates rf magnetic fields on the sample in the test tube. The directional couplers prevent the rf pulses from being transferred to the receiver 1 (2). The signals, which are induced at the coils by the rotating magnetizations of the sample, are led to the receiver 1 (2) and detected.

3.1.2. *Hamiltonian in rotating frame*

The two-qubit Hamiltonian in the laboratory frame is decomposed into three parts as

$$H = H_0 + H_{\rm rf,1} + H_{\rm rf,2}. \tag{45}$$

The first term H_0 describes dynamics of the spins in the static magnetic field and is called the system Hamiltonian after the reference.[7] The rest $H_{\rm rf,i}$ $(i = 1, 2)$ is a Hamiltonian controllable by rf pulses.

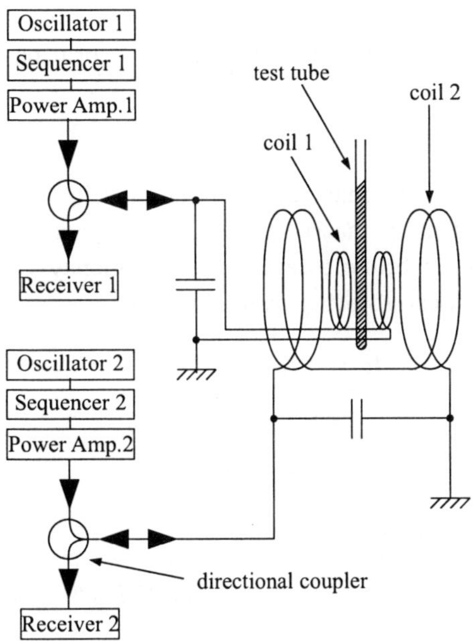

Fig. 3. NMR setup for heteronucleus molecules.

The system Hamiltonian H_0 in the laboratory frame is

$$H_0 = -\hbar\omega_{0,1}\,(I_z \otimes I) - \hbar\omega_{0,2}\,(I \otimes I_z) + \hbar J \sum_{i=x,y,z} (I_i \otimes I_i)\,, \tag{46}$$

where I is the unit matrix of dimension two. The first two terms generate free precession of spins while the third term represents the interaction between the two spins with the coupling strength J. We will not discuss the origin of this interaction. The reader should refer to standard NMR textbooks [9,10] for further details.

The controllable Hamiltonians $H_{\mathrm{rf},i}$ are

$$\begin{aligned} H_{\mathrm{rf},1} &= 2\hbar\omega_{1,1}\cos(\omega_{\mathrm{rf},1}t - \phi_1)\left(I_x \otimes I + \frac{\omega_{0,2}}{\omega_{0,1}} I \otimes I_x\right), \\ H_{\mathrm{rf},2} &= 2\hbar\omega_{1,2}\cos(\omega_{\mathrm{rf},2}t - \phi_2)\left(\frac{\omega_{0,1}}{\omega_{0,2}} I_x \otimes I + I \otimes I_x\right), \end{aligned} \tag{47}$$

where $\omega_{0,2}/\omega_{0,1}$ is the ratio of the gyromagnetic ratios of two spins. The amplitudes $\omega_{1,i}$, the frequencies $\omega_{\mathrm{rf},i}$, and the phases ϕ_i of the rf fields are control parameters. Note that the rf magnetic field is oscillating along the x-axis in the laboratory frame, see Fig. 3.

Let us analyze the Hamiltonian in a rotating frame next. The unitary operator U in Eq. (10) for transformation to a rotating frame is replaced with

$$U(t) = e^{-i\omega_{\mathrm{rot},1} I_z t} \otimes e^{-i\omega_{\mathrm{rot},2} I_z t} \tag{48}$$

if there are two spins. We may choose different values for $\omega_{\mathrm{rot},1}$ and $\omega_{\mathrm{rot},2}$. The transformed density matrix $\tilde{\rho} = U\rho U^\dagger$ represents the state of the system viewed from a frame rotating with frequency $\omega_{\mathrm{rot},i}$ for the spin i. The Hamiltonian in the rotating frame becomes

$$\begin{aligned} \tilde{H} &= UHU^\dagger - i\hbar U \frac{d}{dt} U^\dagger \\ &= \tilde{H}_0 + \tilde{H}_{\mathrm{rf},1} + \tilde{H}_{\mathrm{rf},2}. \end{aligned} \tag{49}$$

The transformed system Hamiltonian is

$$\begin{aligned} \tilde{H}_0 &= UH_0U^\dagger - i\hbar U \frac{d}{dt} U^\dagger \\ &= -\hbar(\omega_{0,1} - \omega_{\mathrm{rot},1})\,(I_z \otimes I) - \hbar(\omega_{0,2} - \omega_{\mathrm{rot},2})\,(I \otimes I_z) \\ &\quad + \hbar J\,(I_z \otimes I_z) + \frac{\hbar J}{2} \begin{pmatrix} 0 & 0 & 0 & 0 \\ 0 & 0 & e^{-i(\omega_{\mathrm{rot},1} - \omega_{\mathrm{rot},2})t} & 0 \\ 0 & e^{i(\omega_{\mathrm{rot},1} - \omega_{\mathrm{rot},2})t} & 0 & 0 \\ 0 & 0 & 0 & 0 \end{pmatrix}. \end{aligned} \tag{50}$$

It is convenient to set the rotation frequencies $\omega_{\mathrm{rot},i}$ equal to the Larmor frequencies $\omega_{0,i}$. Then the first two terms in Eq. (50) vanish. Let $\Delta\omega_0$ be the difference between Larmor frequencies;

$$\Delta\omega_0 \equiv \omega_{0,2} - \omega_{0,1}. \tag{51}$$

The condition $\Delta\omega_0 \gg J$ is always satisfied for heteronucleus molecules, and hence the last term vanishes if it is averaged over the time scale T such that $\Delta\omega_0^{-1} \ll T \ll J^{-1}$. For example, $\Delta\omega_0 \sim 400$ MHz, while $J \sim 200$ Hz for ^{13}C-labeled chloroform at 11 T. Therefore, after setting $\omega_{\mathrm{rot},i} = \omega_{0,i}$, $\tilde{H}_0$ is approximated as

$$\tilde{H}_0 = \hbar J\,(I_z \otimes I_z)\,. \tag{52}$$

Let us consider the controllable Hamiltonians next. They are simplified if we set $\omega_{\mathrm{rot},i} = \omega_{0,i} = \omega_{\mathrm{rf},i}$ as

$$\begin{aligned} \tilde{H}_{\mathrm{rf},1} &= \hbar\omega_{1,1}\left[\cos\phi_1\ (I_x \otimes I) + \sin\phi_1\ (I_y \otimes I)\right], \\ \tilde{H}_{\mathrm{rf},2} &= \hbar\omega_{1,2}\left[\cos\phi_2\ (I \otimes I_x) + \sin\phi_2\ (I \otimes I_y)\right], \end{aligned} \tag{53}$$

after eliminating terms rapidly oscillating with the frequencies $2\omega_{0,i}$ or $\Delta\omega_0$. It is important to note that pulses with the frequency $\omega_{\mathrm{rf},i}$ influence only the spin i but has no effect on the other spin in the rotating frame. This is because the difference $\Delta\omega_0$ in the resonance frequencies is much larger than the inverse of the typical pulse width $\sim 1/(10 \cdot 10^{-6})\ \mathrm{s}^{-1} \sim 100$ kHz. In other words, the pulses with frequency $\omega_{\mathrm{rf},i}$ does not contain the Fourier component which resonates with the other spin. See also § 2.6.4.

In conclusion, the Hamiltonian for heteronucleus molecules in the rotating frame is

$$\begin{aligned}\tilde{H} &= \hbar J\,(I_z \otimes I_z) \\ &+\hbar\,\omega_{1,1}\left[\cos\,\phi_1\ (I_x \otimes I) + \sin\,\phi_1\ (I_y \otimes I)\right] \\ &+\hbar\,\omega_{1,2}\left[\cos\,\phi_2\ (I \otimes I_x) + \sin\,\phi_2\ (I \otimes I_y)\right]\end{aligned} \tag{54}$$

when the condition $\omega_{\mathrm{rot},i} = \omega_{0,i} = \omega_{\mathrm{rf},i}$ is satisfied.

3.2. *Homonucleus molecules*

3.2.1. *Experimental Setup*

A typical NMR setup with homonucleus molecules is shown schematically in Fig. 4. Cytosine solved in D_2O is employed as a typical example of a sample molecule, where two H nucleus spins are the qubits. When $|\Delta\omega_0|$ is small compared to $\omega_{0,i}$, a common resonance circuit and a power amplifier may be employed to control both spins. For example, $\Delta\omega_0 \sim 800$ Hz for cytosine in D_2O when $\omega_{0,i} = 500$ MHz. However, this small difference in $\omega_{0,i}$ still allows us to address each spin individually.

The oscillator 1 (2) generates an electric current oscillating with the radio frequency $\omega_{0,1}$ ($\omega_{0,2}$). The sequencers modulate the oscillating current to generate rf pulses. Typical pulse widths are of the order of 10 μs when addressing the two spins simultaneously and of the order of $1/\Delta\omega_0 \sim 1$ ms when addressing the spins individually. The rf pulses from the two sequencers are mixed and amplified. The subsequent process is similar to the heteronucleus case, except that two spins of homonucleus molecules are accessed by a single set of resonance circuit and receiver.

3.2.2. *Hamiltonian in rotating frame*

We derive the Hamiltonian for homonucleus spins in the rotating frame, following the procedure similar to the heteronucleus case. If the condition $\Delta\omega_0 \gg J$ is satisfied, the system Hamiltonian in the rotating frame takes the form

$$\tilde{H}_0 = \hbar J \,(I_z \otimes I_z) \tag{55}$$

when we set $\omega_{\text{rot},i} = \omega_{0,i} = \omega_{\text{rf},i}$ as before. This coincides with the Hamiltonian (52) for heteronucleus spins. For the case of cytosine in D_2O, we find $\Delta\omega_0 \sim 800$ Hz while $J \sim 7$ Hz and hence the condition $\Delta\omega_0 \gg J$ is satisfied.

The Hamiltonian $\tilde{H}_{\text{rf},i}$ becomes more complicated even after eliminating terms rapidly oscillating with frequencies $2\omega_{0,i}$ or $(\omega_{0,1} + \omega_{0,2})$. We obtain

$$\begin{aligned}
\tilde{H}_{\text{rf},1} &= \hbar\,\omega_{1,1}\left[\cos\phi_1\;(I_x \otimes I) + \sin\phi_1\;(I_y \otimes I)\right] \\
&\qquad + \hbar\,\omega_{1,1}\left[\cos(\Delta\omega_0\,t + \phi_1)\,(I \otimes I_x) + \sin(\Delta\omega_0\,t + \phi_1)\,(I \otimes I_y)\right], \\
\tilde{H}_{\text{rf},2} &= \hbar\,\omega_{1,2}\left[\cos(-\Delta\omega_0\,t + \phi_2)\,(I_x \otimes I) + \sin(-\Delta\omega_0\,t + \phi_2)\,(I_y \otimes I)\right] \\
&\qquad + \hbar\,\omega_{1,2}\left[\cos\phi_2\;(I \otimes I_x) + \sin\phi_2\;(I \otimes I_y)\right],
\end{aligned} \tag{56}$$

where use has been made of the fact that the ratio $\omega_{0,1}/\omega_{0,2}$ of the Larmor frequencies is very close to 1 for homonucleus spins.

If the pulse width is long enough or if $\omega_{1,i} \ll |\Delta\omega_0|$, the terms in Eq. (56) oscillating with frequency $\Delta\omega_0$ can be neglected according to the discussion in § 2.6.4. Then the Hamiltonians Eq. (56) for homonucleus spins assume

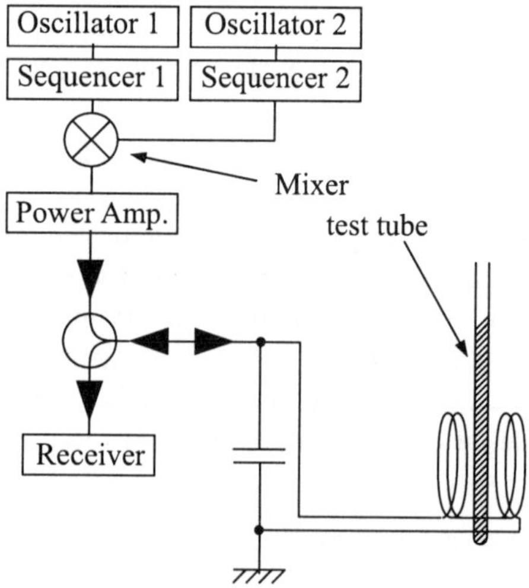

Fig. 4. NMR setup for 2-qubit homonucleus molecules with two oscillators for individual qubits.

the same form as Eq. (53) for heteronucleus spins. At the same time, if the pulse width is short enough compared to $1/J$, the J-coupling effects are negligible during the pulse operations. These observations imply that a pulse sequence designed for heteronucleus spins works also for homonucleus spins if hard pulses are replaced by soft pulses with proper pulse widths.

3.2.3. *Common rotating frame*

In literature, a Hamiltonian

$$\tilde{H}_{0,\mathrm{conv}} = -\hbar\Delta\omega_0 \left(I \otimes I_z\right) + \hbar J \left(I_z \otimes I_z\right) \tag{57}$$

is often used as the system Hamiltonian for homonucleus molecules with two spins in a rotating frame.[13] We call Eq. (57) the conventional Hamiltonian. The above Hamiltonian is derived from the Hamiltonian

$$H_{0,\mathrm{conv}} = -\hbar\omega_{0,1} \left(I_z \otimes I\right) - \hbar\omega_{0,2} \left(I \otimes I_z\right) + \hbar J \left(I_z \otimes I_z\right) \tag{58}$$

in the laboratory frame through a transformation to a common rotating frame. The transformation operator to frames rotating with a common frequency $\omega_{0,1}$ is

$$U = \exp(-i\omega_{0,1}\, I_z t) \otimes \exp(-i\omega_{0,1}\, I_z t), \tag{59}$$

which is obtained from U in Eq. (48) by taking a common frequency $\omega_{\mathrm{rot},1} = \omega_{\mathrm{rot},2} = \omega_{0,1}$. Then the Hamiltonian (46) transforms into

$$\begin{aligned} \tilde{H}_0 &= U H_0 U^\dagger - iU \frac{d}{dt} U^\dagger \\ &= -\Delta\omega_0 \left(I \otimes I_z\right) + J \left(I_z \otimes I_z\right) + \begin{pmatrix} 0 & 0 & 0 & 0 \\ 0 & 0 & \frac{J}{2} & 0 \\ 0 & \frac{J}{2} & 0 & 0 \\ 0 & 0 & 0 & 0 \end{pmatrix}. \end{aligned} \tag{60}$$

The time development operators calculated with Eq. (58) and Eq. (60)

are related as,

$$
\begin{aligned}
&\exp\left(-\frac{i}{\hbar}\int_0^t \tilde{H}_0\,dt\right)\\
&= \begin{pmatrix} 1 & 0 & 0 & 0\\ 0 & f(t) & -g(t)^* & 0\\ 0 & g(t) & f(t)^* & 0\\ 0 & 0 & 0 & 1 \end{pmatrix} \cdot \exp\left(-\frac{i}{\hbar}\int_0^t \tilde{H}_{0,\mathrm{conv}}\,dt\right), \qquad (61)\\
&f(t) = e^{i\Delta\omega_0 t/2}\left(\cos\frac{\Omega t}{2} - i\frac{\Delta\omega_0}{\Omega}\sin\frac{\Omega t}{2}\right),\\
&g(t) = -i\,e^{i\Delta\omega_0 t/2}\,\frac{J}{\Omega}\sin\frac{\Omega t}{2},
\end{aligned}
$$

where $\Omega = \Delta\omega_0\sqrt{1+\epsilon_J^2}$ with $\epsilon_J = J/|\Delta\omega_0|$. Therefore, in the case of $t \lesssim 1/J$ and the weak coupling limit ($\epsilon_J \ll 1$), the time development operator calculated with $\tilde{H}_0$ can be approximated by that with $\tilde{H}_{0,\mathrm{conv}}$. This fact justifies that Eq. (58) can be employed as an approximate Hamiltonian Eq. (60) in the common rotating frame.

Let us consider the gate

$$
U_{\mathrm{E}}(\theta) = \exp\left(-i\theta\ I_z \otimes I_z\right) = \begin{pmatrix} e^{-i\theta/4} & 0 & 0 & 0\\ 0 & e^{i\theta/4} & 0 & 0\\ 0 & 0 & e^{i\theta/4} & 0\\ 0 & 0 & 0 & e^{-i\theta/4} \end{pmatrix} \qquad (62)
$$

to show that the conventional Hamiltonian leads to inconvenient consequences. The two-qubit gate $U_{\mathrm{E}}(\pi)$ has been employed to implement a control-not gate along with one-qubit operations.[1] We consider a gate which is obtained by leaving the system with no applied rf pulses for a period t. If we take our Hamiltonian $\tilde{H}_0$ of Eq. (55), the time development operator is

$$
U_J(t) = \exp\left(-\frac{i}{\hbar}\int_0^t \tilde{H}_0\,dt\right). \qquad (63)
$$

The distance between $U_{\mathrm{E}}(\pi)$ and $U_J(t)$ is easily evaluated as

$$
||U_{\mathrm{E}}(\pi) - U_J(t)|| = 2\sqrt{2}\sqrt{1-\cos\frac{1}{4}(Jt-\pi)}, \qquad (64)
$$

where the norm is defined as $||A|| = \sqrt{\mathrm{tr}(A^\dagger A)}$. The operator $U_J(t)$ coincides with U_{E} at $t = 2\pi/2J$. Note also that the distance $||U_{\mathrm{E}}(\pi) - U_J(t)||$ oscillates slowly in t with the period $T = 8\pi/J$.

If we take the conventional Hamiltonian $\tilde{H}_{0,\mathrm{conv}}$ of Eq. (57), in contrast, the distance between $U_{\mathrm{E}}(\pi)$ and $U_{\mathrm{conv}J}(t)$ is

$$||U_{\mathrm{E}}(\pi) - U_{\mathrm{conv}J}(t)|| = 2\sqrt{2}\sqrt{1 - \cos\left(\frac{\Delta\omega_0 t}{2}\right)\cos\frac{1}{4}(Jt - \pi)}. \quad (65)$$

Therefore, if $\tilde{H}_{0,\mathrm{conv}}$ is employed, the operator $U_{\mathrm{conv}J}(t)$ fails to produce $U_{\mathrm{E}}(\pi)$ even at $t = 2\pi/2J$ in general and the distance $||U_{\mathrm{E}}(\pi) - U_{\mathrm{conv}J}(t)||$ oscillates with the frequency $\Delta\omega_0/2$. Although this inconvenience can be overcome by adjusting the phases of following pulses after $U_{\mathrm{E}}(\pi)$ gates, we always employ the Hamiltonian (55) for its simplicity in this contribution.

3.3. *Unitary operations for two spins*

Any quantum gate required for quantum computations can be decomposed into $U(2)$ gates acting on individual qubits and CNOT gates between a certain pair of qubits.[16]

The CNOT gate is a two-qubit operation in that a qubit, called a target qubit, is flipped only when another qubit, called a control qubit, is in the state $|1\rangle$ while the target qubit is left unchanged when the control qubit is in $|0\rangle$. Let us construct such unitary operations from the Hamiltonian $\tilde{H}$ of Eq. (54) in the rotating frame. Our previous analysis confirms that it suffices to consider only the heteronucleus cases since the pulse sequences for homonucleus molecules are obtained from those for heteronucleus molecules by simply replacing hard pulses with soft pulses.

One-qubit operations that we need are rotations by angle θ around $(\cos\phi, \sin\phi, 0)$ and $(0, 0, 1)$ for each spin. They can be achieved by applying proper rf pulses as illustrated in § 2.6.1. The second term in the Hamiltonian Eq. (54) generates

$$U_{\phi,1}(\theta) = U_\phi(\theta) \otimes I, \quad (66)$$

while the third term generates

$$U_{\phi,2}(\theta) = I \otimes U_\phi(\theta), \quad (67)$$

where the indices 1 and 2 label the spins and $U_\phi(\theta)$ is defined as in Eq. (14). Rotations around $(0, 0, 1)$, namely $U_{z,1}(\theta)$ and $U_{z,2}(\theta)$, can be constructed by employing a composite pulse or by redefining the axes in the rotating frame as explained in § 2.6.1. We assume here that the pulse widths are short enough so that the time-development due to the J-coupling term in $\tilde{H}$ is negligible during the rf pulses. This assumption is safely satisfied for heteronucleus molecules, for which these parameters are on the order of $J \sim 100$ Hz and $\omega_{1,i} \sim 25$ kHz. In the case of homonucleus molecules, they are typically $J \sim 10$ Hz and $\omega_{1,i} \sim 100$ Hz and thus one can still ignore the J-coupling term to a first approximation.

We again denote the operations which rotate the i-th spin around the x-, y-, z-, $-x$-, $-y$- and $-z$-axes by an angle $\pi/2$ as $X_i, Y_i, Z_i, \bar{X}_i, \bar{Y}_i, \bar{Z}_i$, see Eq. (32). Combining these elementary one-qubit operations with the two-qubit operation $U_{\mathrm{E}}(\theta)$ we construct various gates that often appear in quantum algorithms. The first example is the Walsh-Hadamard gate H_2 of two qubits

$$H_2 = H_1 \otimes H_1 = \frac{1}{2}\begin{pmatrix} 1 & 1 & 1 & 1 \\ 1 & -1 & 1 & -1 \\ 1 & 1 & -1 & -1 \\ 1 & -1 & -1 & 1 \end{pmatrix}. \tag{68}$$

It is constructed from H_1 in Eq. (35) as

$$\begin{aligned} H_2 &= (iYZZ) \otimes (iYZZ) \\ &= -Y_1 Z_1 Z_1 Y_2 Z_2 Z_2. \end{aligned} \tag{69}$$

The second example is the CNOT gate, which is constructed by combining $U_{\mathrm{J}}(t)$ of Eq. (63) with one-qubit operations as

$$U_{\mathrm{CNOT12}} = \begin{pmatrix} 1\,0\,0\,0 \\ 0\,1\,0\,0 \\ 0\,0\,0\,1 \\ 0\,0\,1\,0 \end{pmatrix} = e^{i\pi/4} Z_1 \bar{Z}_2 X_2 U_{\mathrm{E}}(\pi) Y_2, \tag{70}$$

where the spin 1 is the control qubit while the spin 2 is the target qubit. Similarly,

$$U_{\mathrm{CNOT21}} = \begin{pmatrix} 1\,0\,0\,0 \\ 0\,0\,0\,1 \\ 0\,0\,1\,0 \\ 0\,1\,0\,0 \end{pmatrix} = e^{i\pi/4} \bar{Z}_1 Z_2 X_1 U_{\mathrm{E}}(\pi) Y_1, \tag{71}$$

where the spin 2 (1) is the control (target) qubit. Note that the overall phase $e^{i\pi/4}$, required to make the gate an element of $SU(4)$, is not observable.

3.4. *Field gradient for two spins*

Here we discuss effects of field gradient on two-spin system. Field gradient modifies the system Hamiltonian Eq. (46) to

$$H_0 = -\hbar\omega_{0,1}(1+\nabla z)I_z \otimes I - \hbar\omega_{0,2}(1+\nabla z)I \otimes I_z + \hbar J \sum_i I_i \otimes I_i, \quad (72)$$

where ∇ is the normalized strength of the field gradient along the z-direction. We assume that the sample extension along the z-axis is $-L_0 < z < L_0$. In the frame rotating with frequency $\omega_{0,i}$ for the i-th spin, the relevant Hamiltonian is

$$\tilde{H}_0 = -\hbar\omega_{0,1}\nabla z\, I_z \otimes I - \hbar\omega_{0,2}\nabla z\, I \otimes I_z + \hbar J I_z \otimes I_z. \quad (73)$$

After application of the field gradient for a time interval τ, the state is unitarily transformed by the operator

$$U_G(z) = \exp[-i(-\nabla\omega_{0,1}zI_z \otimes I - \nabla\omega_{0,2}zI \otimes I_z + JI_z \otimes I_z)\tau]. \quad (74)$$

The density matrix $\tilde{\rho}$ in the rotating frame is a 4×4 Hermitian matrix. After application of the field gradient, the density matrix has z-dependence as

$$\begin{aligned} \tilde{\rho}_G(z) &= U_G(z)\tilde{\rho}U_G^\dagger(z) \\ &= \begin{pmatrix} \rho_{11} & * & * & * \\ * & \rho_{22} & e^{-i\Delta\omega_0\tau\nabla z}\rho_{23} & * \\ * & e^{i\Delta\omega_0\tau\nabla z}\rho_{32} & \rho_{33} & * \\ * & * & * & \rho_{44} \end{pmatrix}, \end{aligned} \quad (75)$$

where $*$ are terms containing $e^{i2\omega_{0,i}\nabla z\tau}$ or $e^{i(\omega_{0,1}+\omega_{0,2})\nabla z\tau}$. These oscillating terms vanish after taking average over the sample coordinate z. A typical field gradient is $\nabla = 10$ mT/m, for example, and its duration is $\tau = 1$ ms. The sample length is $2L_0 = 4$ cm in our case.

In the case of heteronucleus molecules, $\omega_{0,1} - \omega_{0,2}$ and $\omega_{0,i}$ are on the same order of magnitude, and thus the density matrix averaged over the sample is approximated by

$$D_G\tilde{\rho} = \begin{pmatrix} \rho_{11} & 0 & 0 & 0 \\ 0 & \rho_{22} & 0 & 0 \\ 0 & 0 & \rho_{33} & 0 \\ 0 & 0 & 0 & \rho_{44} \end{pmatrix} \quad (76)$$

while for homonucleus molecules, $\tilde{\rho}_G$ takes an approximate form

$$D_G\tilde{\rho} = \begin{pmatrix} \rho_{11} & 0 & 0 & 0 \\ 0 & \rho_{22} & \rho_{23} & 0 \\ 0 & \rho_{32} & \rho_{33} & 0 \\ 0 & 0 & 0 & \rho_{44} \end{pmatrix}. \tag{77}$$

The off-diagonal components remain in the latter case since $|\Delta\omega_0|L\nabla\tau \ll 1$ for a typical choice of the parameters.

Therefore, application of a pulsed field gradient works as a non-unitary transformation which eliminates most (in fact, all for heteronucleus molecules) of off-diagonal elements of the density matrix.

4. Density matrix for two spins

4.1. *Thermal state*

4.1.1. *Heteronucleus molecules*

The density matrix in a thermal equilibrium state is given by (see Eq. (5) for one-qubit molecules)

$$\begin{aligned} \rho_{\text{th}} &= \frac{e^{-(H_0/k_BT)}}{\text{tr}(e^{-(H_0/k_BT)})} \\ &\simeq \frac{1}{4}\left(I\otimes I + \frac{\hbar\omega_{0,1}}{k_BT} I_z \otimes I + \frac{\hbar\omega_{0,2}}{k_BT} I \otimes I_z\right) \\ &= \frac{1}{4} I\otimes I + \frac{\hbar}{8k_BT}\begin{pmatrix} \omega_{0,1}+\omega_{0,2} & 0 & 0 & 0 \\ 0 & \omega_{0,1}-\omega_{0,2} & 0 & 0 \\ 0 & 0 & -\omega_{0,1}+\omega_{0,2} & 0 \\ 0 & 0 & 0 & -\omega_{0,1}-\omega_{0,2} \end{pmatrix}, \end{aligned} \tag{78}$$

where H_0 is given in Eq. (46). We can safely drop the term that contains J since $\omega_{0,i} \gg J$. Note that the matrix ρ_{th} is normalized as tr $\rho_{\text{th}} = 1$.

4.1.2. *Homonucleus Molecules*

In the case of homonucleus molecules, the density matrix is further simplified as

$$\begin{aligned}\rho_{\text{th}} &\simeq \frac{1}{4}I\otimes I+\frac{\hbar\omega_{0,1}}{4k_BT}I_z\otimes I+\frac{\hbar\omega_{0,1}(1+\Delta\omega/\omega_{0,1})}{4k_BT}I\otimes I_z\\ &=\frac{1}{4}I\otimes I+\frac{\hbar\omega_{0,1}}{4k_BT}\begin{pmatrix}1&0&0&0\\0&0&0&0\\0&0&0&0\\0&0&0&-1\end{pmatrix},\end{aligned} \tag{79}$$

where use has been made of the condition $\Delta\omega/\,\omega_{0,1}\ll 1$.

4.2. *Pseudopure state*

The density matrix for n-spin molecules in NMR can be generally written as

$$\rho=\left(\frac{I}{2}\right)^{\otimes n}+\Delta\rho \tag{80}$$

where "$\otimes n$" denotes the n-th tensor power. The first term $(I/2)^{\otimes n}$ represents an isotropic mixed ensemble, in which all spin states appear with equal probability. The second term $\Delta\rho$ represents a deviation from the uniform ensemble. See also Eqs. (78) and (79). A unitary transformation U acts on the density matrix as

$$\begin{aligned}U\rho U^\dagger &= U\left[\left(\frac{I}{2}\right)^{\otimes n}+\Delta\rho\right]U^\dagger\\ &=\left(\frac{I}{2}\right)^{\otimes n}+U\Delta\rho U^\dagger.\end{aligned} \tag{81}$$

This implies that only the deviation from the isotropic ensemble has relevance in time development of the system and the isotropic term $(I/2)^{\otimes n}$ does not contribute to NMR signals.

The mixed state

$$\rho_{\text{pps}}=\left(\frac{I}{2}\right)^{\otimes n}+\alpha\,\text{diag}(1,\underbrace{0,\ldots,0}_{2^n-1}), \tag{82}$$

is effectively equivalent to the pure state $|00\ldots0\rangle$. The real parameter α is of the order of $\hbar\omega_{0,i}/k_BT$. The non-vanishing component is not necessarily the first one. A quantum system whose state is exactly known is in a pure state. In other words, a state is a pure state if and only if the rank of the density matrix is unity. Therefore, we treat the state Eq. (82) as if it were a pure state and thus we call it a *pseudopure state.*

For a quantum computation the system should be initialized to be in a fiducial pure state. However, no unitary transformation $\rho \mapsto U\rho U^\dagger$ changes the rank of ρ, see Eq. (81). Hence it is impossible to get the pseudopure state from a thermal state by a unitary time development. Therefore non-unitary transformations are required to prepare pseudopure states.

4.3. *Initialization*

A naive method to produce a pure state would be to cool the system under consideration. When the thermal energy $k_B T$ becomes much smaller than the energy difference between the ground state and the first excited state, the system is definitely in the ground state. However, this method is not applicable to liquid-state NMR, since it usually works at room temperature and there the thermal energy is much larger than the Zeeman energy of a nucleus. Therefore we need a different method to produce a pseudopure state.

Nonunitary transformations to produce a pseudopure state are classified into three categories: temporal averaging, spatial averaging, and logical labeling. Temporal and spatial averagings are based on the linearity of quantum mechanics. Suppose that there are N initial density matrices $\rho_{I,i}$. Let the same unitary operator U act on them. Then it yields N output density matrices $\rho_{O,i}$ as

$$\rho_{I,i} \xrightarrow{U} \rho_{O,i}. \tag{83}$$

Quantum mechanical linearity guarantees that

$$\sum_i \rho_{O,i} = U\left(\sum_i \rho_{I,i}\right)U^\dagger. \tag{84}$$

If $\sum_i \rho_{I,i}$ is proportional to a pseudopure state ρ_{pps}, then $\sum_i \rho_{O,i}$ is proportional to $U\rho_{\text{pps}}U^\dagger$.

4.3.1. *Temporal averaging*

A suitable pulse sequence in NMR realizes a cyclic permutation of diagonal elements of the thermal density matrix

$$\rho_{\text{th}} = \left(\frac{I}{2}\right)^{\otimes 2} + \text{diag}(a_{11}, a_{22}, a_{33}, a_{44}), \tag{85}$$

see Eq. (78). Two CNOT operations permute the diagonal elements as

$$\begin{aligned}\rho_{\rm th} &\overset{U_{\rm CNOT12}U_{\rm CNOT21}}{\longrightarrow} \rho_1 = \left(\frac{I}{2}\right)^{\otimes 2} + \mathrm{diag}(a_{11}, a_{44}, a_{22}, a_{33}),\\ \rho_{\rm th} &\overset{U_{\rm CNOT21}U_{\rm CNOT12}}{\longrightarrow} \rho_2 = \left(\frac{I}{2}\right)^{\otimes 2} + \mathrm{diag}(a_{11}, a_{33}, a_{44}, a_{22}).\end{aligned} \tag{86}$$

Note that the element a_{11} is left invariant under these transformations. We obtain, after averaging over the density matrices,

$$\begin{aligned}\rho_{\rm pps} &= \frac{1}{3}(\rho_{\rm th} + \rho_1 + \rho_2)\\ &= \left(\frac{I}{2}\right)^{\otimes 2} + \frac{1}{3}(a_{22} + a_{33} + a_{44})\begin{pmatrix}1&0&0&0\\0&1&0&0\\0&0&1&0\\0&0&0&1\end{pmatrix}\\ &\quad + \frac{1}{3}(3a_{11} - a_{22} - a_{33} - a_{44})\begin{pmatrix}1&0&0&0\\0&0&0&0\\0&0&0&0\\0&0&0&0\end{pmatrix}\\ &= \left[1 + \frac{4}{3}(a_{22} + a_{33} + a_{44})\right]\left(\frac{I}{2}\right)^{\otimes 2}\\ &\quad + \frac{1}{3}(3a_{11} - a_{22} - a_{33} - a_{44})\begin{pmatrix}1&0&0&0\\0&0&0&0\\0&0&0&0\\0&0&0&0\end{pmatrix}.\end{aligned} \tag{87}$$

This is a pseudopure state corresponding to a pure state $|00\rangle$. In the case of homonucleus molecules, $(3a_{11} - a_{22} - a_{33} - a_{44})/3 = (2/3)(\hbar\omega_{0,1}/4k_{\rm B}T)$.

In the case of homonucleus molecules (79) we can take temporal average more easily by taking into account the structure of the thermal density matrix

$$\rho_{\rm th} - \left(\frac{I}{2}\right)^{\otimes 2} = \frac{\hbar\omega_{0,1}}{4k_{\rm B}T}(I_z \otimes I + I \otimes I_z) \tag{88}$$

which implies that the thermal state itself may be regarded as a mixture of two states.

From this observation, we find that $\rho_{\rm pps}$ may be constructed with less number of gates, as follows.

$$\rho_{\text{th}} \overset{U_{\text{CNOT12}}}{\longrightarrow} \rho_1 = \left(\frac{I}{2}\right)^{\otimes 2} + \frac{\hbar\omega_{0,1}}{4k_{\text{B}}T}(I_z \otimes I + 2I_z \otimes I_z)$$

$$\rho_{\text{th}} \overset{U_{\text{CNOT21}}}{\longrightarrow} \rho_2 = \left(\frac{I}{2}\right)^{\otimes 2} + \frac{\hbar\omega_{0,1}}{4k_{\text{B}}T}(2I_z \otimes I_z + I \otimes I_z)$$

Note that

$$\frac{1}{3}\left[(I_z \otimes I + I \otimes I_z) + (I_z \otimes I + 2I_z \otimes I_z) + (2I_z \otimes I_z + I \otimes I_z)\right]$$

$$= \frac{1}{3}\begin{pmatrix} 3 & 0 & 0 & 0 \\ 0 & -1 & 0 & 0 \\ 0 & 0 & -1 & 0 \\ 0 & 0 & 0 & -1 \end{pmatrix} = \frac{4}{3}\begin{pmatrix} 1 & 0 & 0 & 0 \\ 0 & 0 & 0 & 0 \\ 0 & 0 & 0 & 0 \\ 0 & 0 & 0 & 0 \end{pmatrix} - \frac{1}{3}I^{\otimes 2}$$

Therefore,

$$\begin{aligned} \rho_{\text{pps}} &= \frac{1}{3}(\rho_{\text{th}} + \rho_1 + \rho_2) \\ &\equiv \frac{4}{3}\frac{\hbar\omega_{0,1}}{4k_{\text{B}}T}\begin{pmatrix} 1 & 0 & 0 & 0 \\ 0 & 0 & 0 & 0 \\ 0 & 0 & 0 & 0 \\ 0 & 0 & 0 & 0 \end{pmatrix} \quad (\text{mod}\, I^{\otimes 2}). \end{aligned} \tag{89}$$

Here we omit irrelevant terms that are proportional to $I^{\otimes 2}$. The above construction is called a product operator approach.[8]

For two-qubit molecules, the number of initial states required to construct a pseudopure state by the product operator approach is three, which is the same as the necessary number required for cyclic permutation approach. However, the product operator approach provides twice larger signal than the cyclic permutation approach. Moreover, its preparation is easier (two CNOT gates in total) than the cyclic permutation approach which demands four CNOT gates. When the number of qubits is more than three, the product operator approach is more advantageous. For example, three initial states are enough for the product operator approach even for three-qubit molecules, while the cyclic permutation approach requires $2^3 - 1 = 7$ initial states.

4.3.2. *Spatial averaging*

Spatial averaging approach employs pulsed field gradients to prepare $\rho_{I,i}$ in Eq. (84).

We consider here homonucleus molecules first. A state

$$\rho_{\mathrm{th}} \overset{D_{G1}U_{\pi/2,2}(\pi/3)}{\longrightarrow} \frac{\hbar\omega_{0,1}}{4k_{\mathrm{B}}T}\left(I_z \otimes I + \frac{1}{2} I \otimes I_z\right) \tag{90}$$

is prepared by a one-qubit operation and the field gradient (spatial averaging). We omit irrelevant terms that are proportional to $(I/2)^{\otimes 2}$ hereafter. Subsequent operations yield

$$\begin{aligned} &\overset{D_{G2}U_{3\pi/2,1}(\pi/4)U_{\mathrm{E}}(\pi)U_{0,1}(\pi/4)}{\longrightarrow} \frac{\hbar\omega_{0,1}}{4k_{\mathrm{B}}T}\left(I_z \otimes I_z + \frac{1}{2} I_z \otimes I + \frac{1}{2} I \otimes I_z\right) \quad (91)\\ &\equiv \frac{\hbar\omega_{0,1}}{4k_{\mathrm{B}}T}\begin{pmatrix} 1 & 0 & 0 & 0 \\ 0 & 0 & 0 & 0 \\ 0 & 0 & 0 & 0 \\ 0 & 0 & 0 & 0 \end{pmatrix} \qquad (\mathrm{mod}\, I^{\otimes 2}). \end{aligned}$$

as promised.

If heteronucleus molecules, such as ^{13}C-labeled chloroform, are employed, the first operation is replaced as

$$D_{G1}U_{\pi/2,2}(\pi/3) \to D_{G1}U_{\pi/2,1}(\eta), \tag{92}$$

where η satisfies $\omega_{0,1}\cos\eta = 2\omega_{0,2}$. Then,

$$\frac{\hbar\omega_{0,1}}{4k_{\mathrm{B}}T} I_z \otimes I + \frac{\hbar\omega_{0,2}}{4k_{\mathrm{B}}T} I \otimes I_z \overset{D_{G1}U_{\pi/2,1}(\eta)}{\longrightarrow} \frac{\hbar\omega_{0,2}}{4k_{\mathrm{B}}T}\left(2I_z \otimes I + I \otimes I_z\right), \tag{93}$$

and the same second operation is applicable to generate a pseudopure state. There is another approach for equalizing populations[17] with

$$D_{\mathrm{eq}} = D_G \bar{X}_1 \bar{X}_2 U_{\mathrm{E}}(\pi/2) Y_1 Y_2 U_{\mathrm{E}}(\pi/2) X_1 X_2. \tag{94}$$

Applying this we obtain

$$\frac{\hbar\omega_{0,1}}{4k_{\mathrm{B}}T} I_z \otimes I + \frac{\hbar\omega_{0,2}}{4k_{\mathrm{B}}T} I \otimes I_z \overset{D_{\mathrm{eq}}}{\longrightarrow} \frac{\hbar(\omega_{0,1}+\omega_{0,2})}{4k_{\mathrm{B}}T}\left(I_z \otimes I + I \otimes I_z\right), \tag{95}$$

after averaging over the sample molecules. The same operations are applicable once the populations are equalized.

General procedures for arbitrary number of qubits are proposed by several groups.[18,19] The method discussed here is a two-qubit version of Sakaguchi, Ozawa and Fukumi[18] which does not require an ancilla spin. In contrast, a method proposed by Sharf, Havel and Cory[19] requires an ancilla spin.

4.3.3. *Logical labeling*

Logical labeling approach[20] to create a pseudopure state is understood by recalling the definition of a "cold" state. At absolute zero temperature, all the spins of a molecule align in the state $|\uparrow\rangle$. Suppose we have a molecule with N spins. If one can rearrange the populations with one- and two-qubit operations so that $N - M$ of N-spins are aligned to $|\uparrow\rangle$ under a certain spin configuration of the other M-spins. In this case, $N - M$-spins can be considered at "0" K and it can be taken as a pseudopure state.

Let us consider a more concrete case of molecules with 3 homonucleus spins. The thermal state of these spins is,

$$\rho_{\text{th}} \propto \text{diag}(3, 1, 1, -1, 1, -1, -1, -3). \tag{96}$$

By using a cyclic permutation of the populations, which is realized with two CNOT gates, ρ_{th} is converted to,

$$\begin{aligned}\text{diag}(3, 1, 1, 1, -1, -1, -1, -3) &= I_z \otimes \text{diag}(3, 1, 1, 1) \\ &\equiv 2I_z \otimes \begin{pmatrix} 1 & 0 & 0 & 0 \\ 0 & 0 & 0 & 0 \\ 0 & 0 & 0 & 0 \\ 0 & 0 & 0 & 0 \end{pmatrix},\end{aligned} \tag{97}$$

where we have dropped $(I/2)^{\otimes n}$ as before. Therefore, when the first spin is $|\uparrow\rangle$, the other two spins are considered at "0" K and thus this subspace is considered to be a pseudopure state.

Suppose we have an N-qubit homonucleus molecule. The maximum number of qubits N_m in a pseudopure state by logical labeling is found as follows. Let N_{th} be the maximum number of equally populated states in thermal equilibrium. Then N_m is given by the integer which does not exceed $\log_2(N_{\text{th}} + 1)$. In the 3-qubit example considered above, $N_{\text{th}} = 3$, namely there are three 1's (or -1's) in the density matrix, see Eq. (96) and hence we find $N_m = 2$.

When N is even, we find $N_{\text{th}} = \binom{N}{N/2} = N!/(N/2)!$ since N_{th} is the number of states in which $N/2$ $|\uparrow\rangle$'s and $N/2$ $|\downarrow\rangle$'s are involved. On the other hand, when N is odd, $N_{\text{th}} = \binom{N}{(N+1)/2}$ because N_{th} is the number of states in which $(N+1)/2$ of $|\uparrow\rangle$ and $(N-1)/2$ of $|\downarrow\rangle$ are involved.

We do not discuss the logical labeling approach in the rest of this article.

4.4. *Quantum state tomography*

The density matrix $\tilde{\rho}$ of two-qubit molecules in a rotating frame is parametrized as

$$\tilde{\rho} = \frac{I}{2} \otimes \frac{I}{2} + \begin{pmatrix} a_{11} & a_{12} + ib_{12} & a_{13} + ib_{13} & a_{14} + ib_{14} \\ a_{12} - ib_{12} & a_{22} & a_{23} + ib_{23} & a_{24} + ib_{24} \\ a_{13} - ib_{13} & a_{23} - ib_{23} & a_{33} & a_{34} + ib_{34} \\ a_{14} - ib_{14} & a_{24} - ib_{24} & a_{34} - ib_{34} & a_{44} \end{pmatrix}. \tag{98}$$

Note that $I/2 \otimes I/2$ is put for normalization and does not contribute to NMR signals. The number of independent parameters is $16 - 1 = 15$, because the constraint $\mathrm{tr}\tilde{\rho} = 1$ introduces a relation among $\{a_{ii}\}$. This is also understood as follows. The density matrix of two-spin molecules is expressed as

$$\tilde{\rho} = \sum_{i,j=0,x,y,z} c_{ij} I_i \otimes I_j, \tag{99}$$

where $I_0 \equiv I/2$. Since $c_{00} = 1$ for normalization, the number of free parameters is $4 \times 4 - 1 = 15$.

By measuring all a_{ij} and b_{ij}, we know the state completely: this is called *quantum state tomography*, see also § 2.4. A state of a multi-qubit system develops in time driven by J-couplings among spins. Therefore, the actual final state after operation of a quantum algorithm (63) in the rotating frame is

$$\tilde{\rho}(t) = U_J(t)\tilde{\rho}U_J^\dagger(t). \tag{100}$$

The density matrix $\rho(t)$ in the laboratory frame is

$$\begin{aligned} \rho(t) &= U^\dagger \tilde{\rho}(t) U \\ &= \exp\left[i\omega_{0,1}\, t(I_z \otimes I) + i\omega_{0,2}\, t(I \otimes I_z)\right] \tilde{\rho}(t) \\ &\quad \cdot \exp\left[-i\omega_{0,1}\, t(I_z \otimes I) - i\omega_{0,2}\, t(I \otimes I_z)\right]. \end{aligned} \tag{101}$$

Then the x-component of the magnetization in the laboratory frame is

$$\begin{aligned}
&\mathrm{tr}\left[(\sigma_x \otimes I + I \otimes \sigma_x)\rho(t)\right] \\
&= \Big[a_{12}\cos\left(\omega_{0,2}-\frac{J}{2}\right)t + a_{34}\cos\left(\omega_{0,2}+\frac{J}{2}\right)t \\
&\quad + a_{13}\cos\left(\omega_{0,1}-\frac{J}{2}\right)t + a_{24}\cos\left(\omega_{0,1}+\frac{J}{2}\right)t\Big] \\
&\quad - \Big[b_{12}\sin\left(\omega_{0,2}-\frac{J}{2}\right)t + b_{34}\sin\left(\omega_{0,2}-\frac{J}{2}\right)t \\
&\quad + b_{13}\sin\left(\omega_{0,1}-\frac{J}{2}\right)t + b_{24}\sin\left(\omega_{0,1}-\frac{J}{2}\right)t\Big].
\end{aligned} \tag{102}$$

Therefore $a_{12}, a_{34}, a_{13}, a_{24}$ are measured from the real part of the spectrum while $b_{12}, b_{34}, b_{13}, b_{24}$ from the imaginary part. See, § 2.5.

Suppose next that a $\pi/2$-pulse Y_1 around the y-axis in the rotating frame is applied to the spin 1. The density matrix after this operation is

$$Y_1 \tilde{\rho} Y_1^\dagger. \tag{103}$$

The x-projections in the laboratory frame after this $\pi/2$-pulse is

$$\begin{aligned}
&\Big[\left(\frac{a_{12}}{2}-\frac{a_{14}}{2}-\frac{a_{23}}{2}+\frac{a_{34}}{2}\right)\cos\left(\omega_{0,2}-\frac{J}{2}\right)t \\
&+\left(\frac{a_{12}}{2}+\frac{a_{14}}{2}+\frac{a_{23}}{2}+\frac{a_{34}}{2}\right)\cos\left(\omega_{0,2}+\frac{J}{2}\right)t \\
&+\left(\frac{a_{11}}{2}-\frac{a_{33}}{2}\right)\cos\left(\omega_{0,1}-\frac{J}{2}\right)t+\left(\frac{a_{22}}{2}-\frac{a_{44}}{2}\right)\cos\left(\omega_{0,1}+\frac{J}{2}\right)t\Big] \\
+&\Big[-\left(\frac{b_{12}}{2}-\frac{b_{14}}{2}+\frac{b_{23}}{2}+\frac{b_{34}}{2}\right)\sin\left(\omega_{0,2}-\frac{J}{2}\right)t \\
&+b_{24}\sin\left(\omega_{0,2}+\frac{J}{2}\right)t-b_{13}\sin\left(\omega_{0,1}-\frac{J}{2}\right)t \\
&+\left(-\frac{b_{12}}{2}-\frac{b_{14}}{2}+\frac{b_{23}}{2}-\frac{b_{34}}{2}\right)\sin\left(\omega_{0,1}+\frac{J}{2}\right)t\Big].
\end{aligned} \tag{104}$$

We extract from (104) the information

$$a_{11}-a_{33}, \quad a_{22}-a_{44}, \quad a_{14}+a_{23}, \quad b_{14}-b_{23}.$$

Similary, measurement after a Y_2-pulse provides the information

$$a_{11}-a_{22}, \quad a_{33}-a_{44}, \quad a_{14}+a_{23}, \quad b_{14}+b_{23}$$

while measurement after a X_1-pulse provides the inforamtion

$$b_{14} + b_{23}, \quad a_{11} - a_{33}, \quad a_{22} - a_{44}, \quad a_{14} - a_{23}.$$

Therefore, all a_{ij} and b_{ij} can be determined experimentally. Note that there are other combinations of measurements to yield $\{a_{ij}\}$ and $\{b_{ij}\}$.

Let us consider a simple case in which $\tilde{\rho} = \mathrm{diag}(a_{11}, a_{22}, a_{33}, a_{44})$. Application of a $\pi/2$-pulse Y_1 leads to the following x-projection:

$$\begin{aligned} \mathrm{tr}\left[(\sigma_x \otimes I)\, Y_1 \tilde{\rho} Y_1^\dagger\right] = \frac{1}{2}\Bigg[& a_{11} \cos\left(\omega_{0,1} - \frac{J}{2}\right) t + a_{22} \cos\left(\omega_{0,1} + \frac{J}{2}\right) t \\ & - a_{33} \cos\left(\omega_{0,1} - \frac{J}{2}\right) t - a_{44} \cos\left(\omega_{0,1} + \frac{J}{2}\right) t\Bigg]. \end{aligned} \tag{105}$$

Since not few quantum computations result in a density matrix in which only one diagonal element is 1 while all the other elements being 0, the above method provides a quick way to confirm the result of computation. Equation (105) tells us that the state of the spin 1 can be read from the signature in front of a cos function; the signature + (−) corresponds to $|\uparrow\rangle = |0\rangle$ ($|\downarrow\rangle = |1\rangle$). The state of the spin 2 is read from the peak position in the spectrum; a larger (smaller) frequency shift corresponds to $|\uparrow\rangle$ ($|\downarrow\rangle$) for the spin 2, assuming $J > 0$.

5. Quantum computation

We have demonstrated so far that we can perform one- and two-qubit operations and measure the quantum states of the system using Hamiltonian formulation. We are now ready to perform quantum computations with two-qubit molecules. We examine here two most popular quantum algorithms; the Deutsch-Jozsa (DJ) algorithm and Grover's database search algorithm. First, we will briefly review these algorithms and show the results of our experiments which reproduce the results reported in literature. Then, we will introduce our own contributions.

5.1. *Deutsch-Jozsa algorithm*

5.1.1. *Background*

The DJ algorithm is a quantum algorithm proposed by Deutsch and Jozsa in 1992.[21] It is one of the first algorithms which take advantage of quantum nature, superposition and interference, in computation.

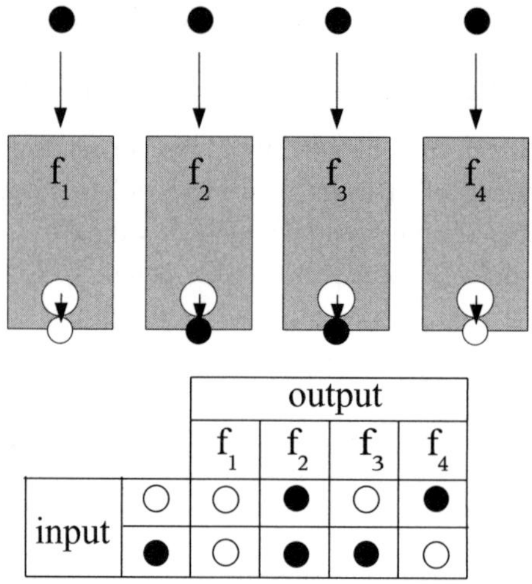

Fig. 5. Deutsch-Jozsa problem for one input variable. The task is to determine whether unknown f_k is constant (f_1 or f_2) or balanced (f_3 or f_4).

Suppose there is an input of one bit, 0 or 1, as shown in Fig. 5. The white ball represents 0 and the black ball represents 1 for example. There are four boxes corresponding to four distinct functions,

$$\begin{aligned} f_1(0) &= 0, & f_1(1) &= 0, \\ f_2(0) &= 1, & f_2(1) &= 1, \\ f_3(0) &= 0, & f_3(1) &= 1, \\ f_4(0) &= 1, & f_4(1) &= 0. \end{aligned}$$

For example, when a black ball ($= 1$) is thrown into the box 1 (function f_1), a white ball ($= 0$) comes out. The task is to determine whether unknown f_k is constant (f_1 or f_2) or balanced (f_3 or f_4, for which half of the output is 0 and the rest is 1) by throwing balls into the unknown box as few times as possible. This task is achieved, quantum mechanically, by throwing a single ball which is a superposition of white and black balls.

In a general DJ problem, we consider a function which has n bits input and one bit output

$$f : \{0,1\}^n \to \{0,1\}; \quad (x_1, x_2, \ldots, x_n) \mapsto x. \tag{106}$$

If $f(x_1, x_2, \ldots, x_n) = 0$ or $f(x_1, x_2, \ldots, x_n) = 1$ for all $(x_1, x_2, \ldots, x_n) \in$

$\{0,1\}^n$, the function f is called constant. If $f(x_1, x_2, \ldots, x_n) = 0$ for half of $\{0,1\}^n$ and $f(x_1, x_2, \ldots, x_n) = 1$ for the rest, the function f is called "balanced". Note that not all functions $f : \{0,1\}^n \to \{0,1\}$ are classified into these two classes. Our task is to determine whether f is constant or balanced assuming that f belongs to one of these two classes. It takes at least $2^{n-1}+1$ steps to tell if a given f is balanced or constant when a classical algorithm is employed. If, instead, the DJ quantum algorithm is employed, we need to evaluate the function once for all. The DJ algorithm is shown in a quantum circuit form in Fig. 6. The XOR (exclusive-OR) operation can be realized with U_{CNOT12} and U_{CNOT21} as shown in Fig. 7.

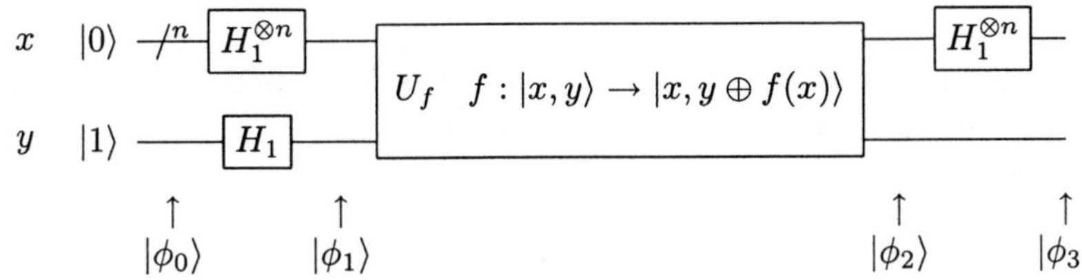

Fig. 6. A quantum circuit implementing the DJ algorithm. The symbol $/^n$ represents a set of n qubits. H_1 is the Hadamard gate. The symbol $\oplus$ is the XOR (exclusive-OR) operation.

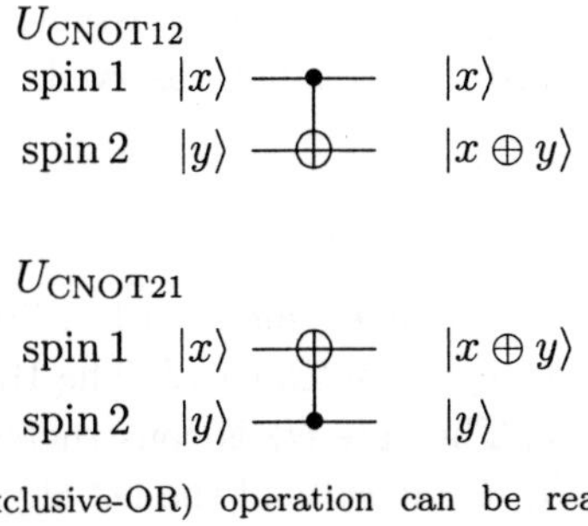

Fig. 7. The XOR (exclusive-OR) operation can be realized with U_{CNOT12} and U_{CNOT21}.

The quantum state develops at each step in Fig. 6 as[1]

$$
\begin{aligned}
|\phi_0\rangle &= |0\rangle^{\otimes n}\,|1\rangle \\
|\phi_1\rangle &= \sum_{x\in\{0,1\}^n} \frac{|x\rangle}{\sqrt{2^n}}\Big(\frac{|0\rangle - |1\rangle}{\sqrt{2}}\Big) \\
|\phi_2\rangle &= \sum_{x\in\{0,1\}^n} \frac{|x\rangle}{\sqrt{2^n}}\Big((-1)^{f(x)}\frac{|0\rangle - |1\rangle}{\sqrt{2}}\Big) \\
|\phi_3\rangle &= \sum_{z\in\{0,1\}^n}\sum_{x\in\{0,1\}^n} \frac{(-1)^{x\cdot z}\,|z\rangle}{2^n}\Big((-1)^{f(x)}\frac{|0\rangle - |1\rangle}{\sqrt{2}}\Big) \\
&= \sum_{z\in\{0,1\}^n}\sum_{x\in\{0,1\}^n} \frac{(-1)^{x\cdot z+f(x)}\,|z\rangle}{2^n}\Big(\frac{|0\rangle - |1\rangle}{\sqrt{2}}\Big).
\end{aligned}
$$

The identity $H^{\otimes n}\,|x\rangle = \sum_z (-1)^{x\cdot z}\,|z\rangle/\sqrt{2^n}$ has been used in evaluating $|\phi_3\rangle$. The coefficient of $|0\rangle^{\otimes n}\,(\,|0\rangle - |1\rangle\,)/\sqrt{2}$ in $|\phi_3\rangle$ is

$$\sum_{x\in\{0,1\}^n} \frac{(-1)^{f(x)}}{2^n}. \tag{107}$$

If f is constant, namely, if $f \equiv 0$ or $f \equiv 1$, Eq. (107) becomes ± 1. On the other hand, if f is balanced, the coefficient vanishes since

$$
\begin{aligned}
\sum_{x\in\{0,1\}^n} \frac{(-1)^{f(x)}}{2^n} &= \sum_{x\in f^{-1}(0)} \frac{(-1)^0}{2^n} + \sum_{x\in f^{-1}(1)} \frac{(-1)^1}{2^n} \\
&= \frac{1}{2} - \frac{1}{2} = 0
\end{aligned} \tag{108}
$$

due to interference among states. Therefore, if the probability to observe the state $|0,0,\ldots,0\rangle$ is 1, then f must be constant. The function f must be balanced if the state $|0,0,\ldots,0\rangle$ is not observed.

5.1.2. *Implementation*

The DJ algorithm (Fig. 5) is implemented with a liquid-state NMR. The NMR pulse sequences are shown in Table 1.[8] The Hadamard gate is constructed with a $\pi/2$-pulse. The gate U_f is constructed with several pulses and the J-coupling, i.e. $U_{\rm E}(\theta)$. Results of the DJ algorithm is summarized in Table 2.

Table 1. The pulse sequences for the DJ algorithm for one-bit functions that produce the unitary matrices in the right column.

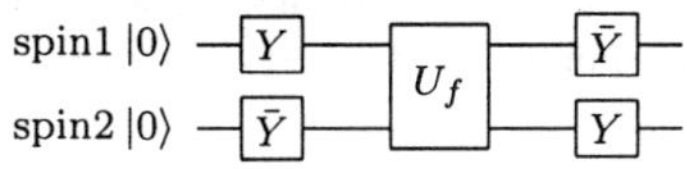

	U_f	
$f_1(x) \equiv 0$	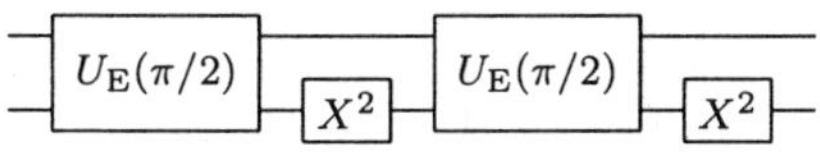	$-\begin{pmatrix} 1&0&0&0 \\ 0&1&0&0 \\ 0&0&1&0 \\ 0&0&0&1 \end{pmatrix}$
$f_2(x) \equiv 1$	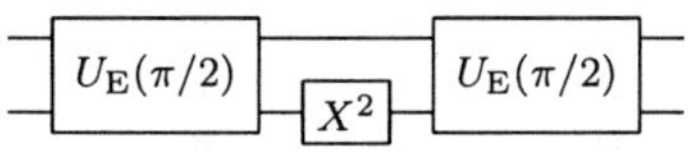	$-\begin{pmatrix} 0&i&0&0 \\ i&0&0&0 \\ 0&0&0&i \\ 0&0&i&0 \end{pmatrix}$
$f_3(x) = x$	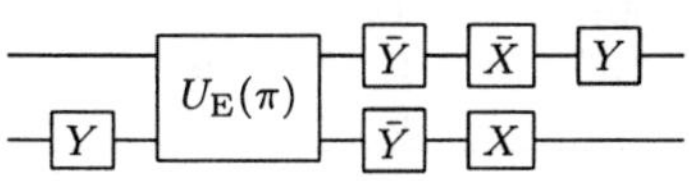	$-(-1)^{3/4}\begin{pmatrix} 1&0&0&0 \\ 0&1&0&0 \\ 0&0&0&1 \\ 0&0&1&0 \end{pmatrix}$
$f_4(x) = 1 - x$	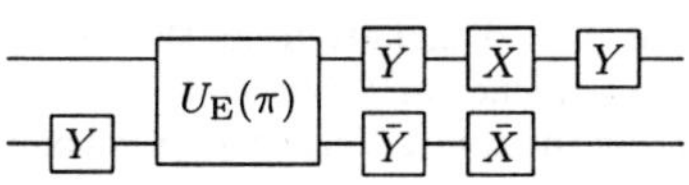	$(-1)^{1/4}\begin{pmatrix} 0&1&0&0 \\ 1&0&0&0 \\ 0&0&1&0 \\ 0&0&0&1 \end{pmatrix}$

5.1.3. *Experiments*

Data were taken at room temperature with a JEOL ECA-500 spectrometer, whose hydrogen Larmor frequency is approximately 500 MHz.[22] We used 0.6 mL, 23 mM sample of cytosine in D_2O.[23] Spins of two hydrogen nuclei in a cytosine molecule work as qubits. The measured coupling strength is $J/2\pi = 7.1$ Hz and the frequency difference is $\Delta\omega/2\pi = 765.0$ Hz. The transverse relaxation time T_2 is ~ 1 s for the both hydrogen nuclei and the longitudinal relaxation time T_1 is ~ 7 s.

It is known that the initial spin state to execute the DJ algorithm may be a thermal state.[24] The density matrix becomes diagonal after the DJ algorithm is executed. Therefore we need to apply one pulse for quantum state tomography as shown in Eq. (105). Expected signals are summarized in Table 3.

Our results of quantum computations with cytosine in D_2O are summarized in Fig. 8. The figures show Fourier transformed spectra of signals from the spin 1. Since the initial state is in thermal equilibrium, molecular spins

Table 2. Inputs and outputs of the Deutsch-Jozsa algorithm. The algorithm is carried out with the pulse sequences given in Table 1. $|x\rangle$ is the state of the spin 1 and $|y\rangle$ is the state of the spin 2. In can be seen that the outputs of the spin 2 gives $f(x) \oplus y$. The phase of states is not considered.

function	input $\|xy\rangle$	$f_i(x)$	$y \oplus f_i(x)$	output $\|xy\rangle$
$f_1(x) = 0$	$\|00\rangle$	0	0	$\|00\rangle$
	$\|01\rangle$	0	1	$\|01\rangle$
	$\|10\rangle$	0	0	$\|10\rangle$
	$\|11\rangle$	0	1	$\|11\rangle$
$f_2(x) = 1$	$\|00\rangle$	1	1	$\|01\rangle$
	$\|01\rangle$	1	0	$\|00\rangle$
	$\|10\rangle$	1	1	$\|11\rangle$
	$\|11\rangle$	1	0	$\|10\rangle$
$f_3(x) = x$	$\|00\rangle$	0	0	$\|00\rangle$
	$\|01\rangle$	0	1	$\|01\rangle$
	$\|10\rangle$	1	1	$\|11\rangle$
	$\|11\rangle$	1	0	$\|10\rangle$
$f_4(x) = \mathrm{NOT}(x)$	$\|00\rangle$	1	1	$\|01\rangle$
	$\|01\rangle$	1	0	$\|00\rangle$
	$\|10\rangle$	0	0	$\|10\rangle$
	$\|11\rangle$	0	1	$\|11\rangle$

Table 3. The expected signals from the resulting state of the pulse sequence in Table 1 followed by the reading pulse Y_1 on the spin 1. The effect of relaxation is ignored.

function	expected signal
f_1	$\cos(\omega_{0,1} - J/2)t + \cos(\omega_{0,1} + J/2)t$
f_2	$\cos(\omega_{0,1} - J/2)t + \cos(\omega_{0,1} + J/2)t$
f_3	$-\cos(\omega_{0,1} - J/2)t + \cos(\omega_{0,1} + J/2)t$
f_4	$-\cos(\omega_{0,1} - J/2)t + \cos(\omega_{0,1} + J/2)t$

are mixture of four states $|00\rangle$, $|01\rangle$, $|10\rangle$, and $|11\rangle$. When the initial state of the spin 2 is $|1\rangle$, the algorithm fails to distinguish a constant function from a balanced function. In contrast, it works regardless of the state of the spin 1.[24] Peaks with a larger frequency shift (the left peaks on each curve) in Fig. 8 correspond to an initial state in which the spin 2 is in the state $|0\rangle$ and successfully tell if f_i is constant or balanced. On the other hand, the peaks with a smaller frequency shift (the right peaks) correspond to the initial state in which the spin 2 is $|1\rangle$ and hence fails.

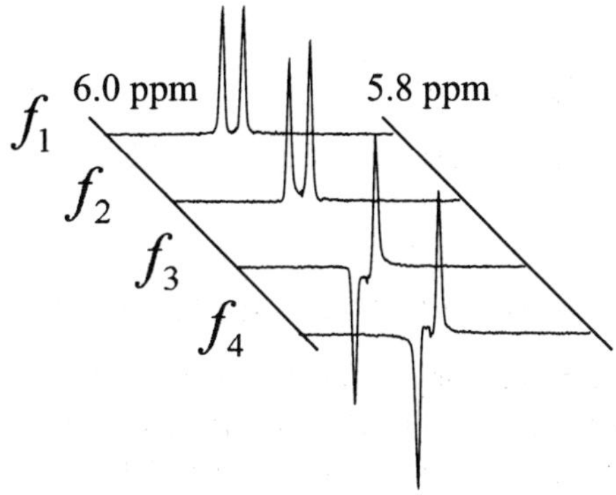

Fig. 8. Spectra of signals from the spin 1 after application of pulse sequences of the Deutsch-Jozsa algorithm and the reading pulse. The peaks with a larger frequency shift (the left peaks) are signals that result from the initial spin 2 state = $|0\rangle$. They tell if the function f_i is constant or balanced from the signature of the peak. The peaks are positive for f_1 and f_2 (constant) and are negative for f_3 and f_4 (balanced).

5.2. *Field inhomogeneity compensation*

5.2.1. *Application of π-pulse pairs during J-coupling operation*

Molecules in NMR quantum computing are often under the influence of field inhomogeneity. This may cause an error in realization of $U_{\rm E}$.

It is well known that an undesired effect caused by field inhomogeneity can be compensated by a series of hard π-pulse pairs, where the width of each pulse is on the order of 10 μs. The best known example may be the CPMG (Carr-Purcell-Meiboom-Gill) pulse sequence.[9] By applying this technique to $U_{\rm E}$, we replace $U_J(t)$ with

$$U_{J*}(t) = U_J(t/2) - \pi - U_J(t/2) - \pi, \tag{109}$$

where π denotes a hard π-pulse around the x-axis of one of the two spins, for example the spin 1, and time flows from the left to the right in the right hand side. According to Eq. (56), the above operation is described as

$$U_{J*}(t) = U_{\pi,1}^{(2)} U_J(t/2) U_{\pi,1}^{(1)} U_J(t/2), \tag{110}$$

where

$$
\begin{aligned}
U_{\pi,1}^{(1)} &= \exp\left(-i\pi I_x \otimes I\right) \\
&\quad \times \exp\left[-i\pi(\cos(\Delta\omega t/2)(I \otimes I_x) + \sin(\Delta\omega t/2)(I \otimes I_y))\right], \\
U_{\pi,1}^{(2)} &= \exp\left(-i\pi I_x \otimes I\right) \\
&\quad \times \exp\left[-i\pi(\cos \Delta\omega t(I \otimes I_x) + \sin \Delta\omega t(I \otimes I_y))\right].
\end{aligned}
\tag{111}
$$

We assume that the π-pulse is around the x-axis of the spin 1 in its rotating frame and is instantaneous. Thus the rotation axis for the spin 2 turns with the angular velocity $\Delta\omega$ during $U_J(t/2)$. The distance between $U_{J*}(t)$ and $U_{\mathrm{E}}(\pi)$ is evaluated as before

$$
\|U_{\mathrm{E}}(\pi) - U_{J*}(t)\| = 2\sqrt{2}\sqrt{1 - \cos\left(\frac{\Delta\omega t}{2}\right)\cos\frac{1}{4}\left(Jt - \pi\right)}. \tag{112}
$$

The distance does not vanish for any t since two conditions $\cos\left(\Delta\omega t/2\right) = \pm 1$ and $\cos\frac{1}{4}\left(Jt - \pi\right) = \pm 1$ are not simultaneously satisfied in general. Moreover, the distance oscillates rapidly in t through $\cos(\Delta\omega t/2)$.

5.2.2. *Pseudopure state by spatial labeling*

We applied a π-pulse pair during the two-qubit entangling operation $U_{\mathrm{E}}(\pi)$ in the pulse sequence which creates the pseudopure state $|00\rangle$ by spatial labeling, which was discussed in § 4.3.2. The entangling operation is realized by turning off rf pulses for a specified time.

The experimental results are summarized in Fig. 9. The spectra in the left panel were obtained without a π-pulse pair. We should observe, in principle, only a single peak with a larger frequency shift, namely a signal from molecules in the state $|00\rangle$. A small peak at frequency with smaller frequency shift indicates the spurious state $|01\rangle$.

Those in the right panel were obtained with a π-pulse pair applied during the entangling operation. We see that they are very sensitive to the entangling operation time t. When t was set at the correct value 70.3 ms, the spectrum exhibited a sharp peak at a frequency with larger shift and a small peak at a frequency with smaller shift. The sharpest peak obtained with the right value of t was sharper than those in the left panel. This result implies that a π-pulse pair improves quality of the pseudopure state. However, even a small deviation of t from the correct value distorts the spectrum considerably. The spectrum with $t = 69.7$ ms, for example, indicates that the created state is actually $|01\rangle$, not the desired $|00\rangle$.

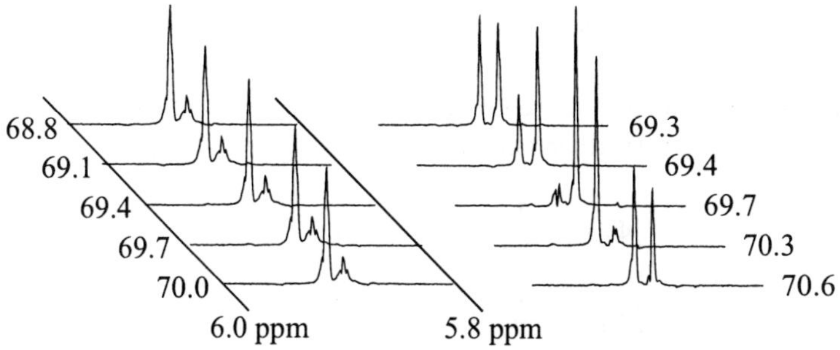

Fig. 9. The spectra of the spin 1 in the pseudopure state. Spectra in the left panel were obtained without π-pulse pairs while the spectra in the right panel were obtained with π-pulse pairs applied during the two-qubit entanglement operation. The entangling time was varied from 68.8 ms to 70.0 ms in the left panel while from 69.3 ms to 70.6 ms in the right panel. The peak with a larger frequency shift (the left peak on each curve) corresponds to the state $|00\rangle$ while the peak with a smaller frequency shift (the right peak) corresponds to the state $|01\rangle$. The latter peak indicates an error in the pseudopure state preparation. The spectra in the right panel are sensitive to variation of the entangling time.

5.3. *Optimal Implementation of Two-Qubit algorithms*

We need to find a control function $\gamma(t)$ in Eq. (1) to implement a quantum algorithm U_{alg} so that

$$U[\gamma(t)] = U_{\text{alg}}. \tag{113}$$

It is shown[25] that U_{alg} can be decomposed into three matrices as

$$U_{\text{alg}} = k_2 h k_1, \tag{114}$$

where $k_1, k_2 \in \mathrm{SU}(2) \otimes \mathrm{SU}(2)$ are SU(4) transformations written as a tensor product of two one-qubit operations,

$$h(t_x, t_y, t_z) = \prod_{i=x,y,z} U_{ii}(t_i). \tag{115}$$

Here

$$\begin{aligned} U_{xx}(t) &= e^{-iJtI_x \otimes I_x} = Y_1 Y_2 U_J(t) \bar{Y}_2 \bar{Y}_1 \\ U_{yy}(t) &= e^{-iJtI_y \otimes I_y} = X_1 X_2 U_J(t) \bar{X}_2 \bar{X}_1 \\ U_{zz}(t) &= U_J(t), \end{aligned} \tag{116}$$

are two-qubit unitary transformations which cannot be decomposed into a tensor product of one-qubit unitary transformations in general. The decomposition (114) is called a Cartan decomposition. Since the execution time of k_i is much shorter than that of $U_J(t_i)$, the total execution time is approximated as $T = \sum_{i=x,y,z} t_i$.

It is possible to explicitly write down the Cartan decomposition of any $U \in \mathrm{SU}(4)$.[25,26] Let us introduce the "magic" basis[27]

$$
\begin{aligned}
|\Psi_0\rangle &= \frac{1}{\sqrt{2}}(|00\rangle + |11\rangle), \\
|\Psi_1\rangle &= \frac{i}{\sqrt{2}}(|01\rangle + |10\rangle), \\
|\Psi_2\rangle &= \frac{1}{\sqrt{2}}(|01\rangle - |10\rangle), \\
|\Psi_3\rangle &= \frac{i}{\sqrt{2}}(|00\rangle - |11\rangle).
\end{aligned} \tag{117}
$$

Under the basis change from the binary basis $|00\rangle, |01\rangle, |10\rangle, |11\rangle$ to the magic basis, a matrix U transforms as $U \to U_B \equiv Q^\dagger U Q$, where

$$
Q = \frac{1}{\sqrt{2}} \begin{pmatrix} 1 & 0 & 0 & i \\ 0 & i & 1 & 0 \\ 0 & i & -1 & 0 \\ 1 & 0 & 0 & -i \end{pmatrix}. \tag{118}
$$

The matrix Q defines an isomorphism between $K = \mathrm{SU}(2) \otimes \mathrm{SU}(2)$ and $\mathrm{SO}(4)$ by $Q^\dagger k_i Q \in \mathrm{SO}(4)$ and is used to classify two-qubit gates.[27,29] Some examples are shown in Table 4. Moreover, Q diagonalizes elements of the Cartan subgroup, viz $Q^\dagger h Q = \mathrm{diag}(e^{i\theta_0}, e^{i\theta_1}, e^{i\theta_2}, e^{i\theta_3})$. Useful examples are found in Table 5.

Table 4. Typical examples of $Q^\dagger k Q$. $U_i(\theta, \phi, \xi)$ is a matrix representation of a rotation by an angle ξ around the axis $(\sin\theta\cos\phi, \sin\theta\sin\phi, \cos\theta)$ of the spin i.

k	$Q^\dagger k Q$
$U_1(\theta, \phi, \xi)$	$\begin{pmatrix} \cos\frac{\xi}{2} & \cos\phi\sin\theta\sin\frac{\xi}{2} & \sin\phi\sin\theta\sin\frac{\xi}{2} & \cos\theta\sin\frac{\xi}{2} \\ -\cos\phi\sin\theta\sin\frac{\xi}{2} & \cos\frac{\xi}{2} & -\cos\theta\sin\frac{\xi}{2} & \sin\phi\sin\theta\sin\frac{\xi}{2} \\ -\sin\phi\sin\theta\sin\frac{\xi}{2} & \cos\theta\sin\frac{\xi}{2} & \cos\frac{\xi}{2} & -\cos\phi\sin\theta\sin\frac{\xi}{2} \\ -\cos\theta\sin\frac{\xi}{2} & -\sin\phi\sin\theta\sin\frac{\xi}{2} & \cos\phi\sin\theta\sin\frac{\xi}{2} & \cos\frac{\xi}{2} \end{pmatrix}$
$U_2(\theta, \phi, \xi)$	$\begin{pmatrix} \cos\frac{\xi}{2} & \cos\phi\sin\theta\sin\frac{\xi}{2} & -\sin\phi\sin\theta\sin\frac{\xi}{2} & \cos\theta\sin\frac{\xi}{2} \\ -\cos\phi\sin\theta\sin\frac{\xi}{2} & \cos\frac{\xi}{2} & \cos\theta\sin\frac{\xi}{2} & \sin\phi\sin\theta\sin\frac{\xi}{2} \\ \sin\phi\sin\theta\sin\frac{\xi}{2} & -\cos\theta\sin\frac{\xi}{2} & \cos\frac{\xi}{2} & \cos\phi\sin\theta\sin\frac{\xi}{2} \\ -\cos\theta\sin\frac{\xi}{2} & -\sin\phi\sin\theta\sin\frac{\xi}{2} & -\cos\phi\sin\theta\sin\frac{\xi}{2} & \cos\frac{\xi}{2} \end{pmatrix}$

Table 5. Typical examples of $Q^\dagger hQ$.

h	$Q^\dagger hQ$
$U_{xx}(2\pi/2J) = e^{-i\pi I_x \otimes I_x}$	$e^{-i\pi/4}\begin{pmatrix} 1&0&0&0 \\ 0&1&0&0 \\ 0&0&i&0 \\ 0&0&0&i \end{pmatrix}$
$U_{yy}(2\pi/2J) = e^{-i\pi I_y \otimes I_y}$	$e^{-i\pi/4}\begin{pmatrix} 1&0&0&0 \\ 0&i&0&0 \\ 0&0&1&0 \\ 0&0&0&i \end{pmatrix}$
$U_{zz}(2\pi/2J) = e^{-i\pi I_z \otimes I_z}$	$e^{-i\pi/4}\begin{pmatrix} 1&0&0&0 \\ 0&i&0&0 \\ 0&0&i&0 \\ 0&0&0&1 \end{pmatrix}$

Let U be a matrix to be decomposed. The matrix U in the magic base takes the form

$$U_B = Q^\dagger U Q. \tag{119}$$

If $U_B \in \mathrm{SO}(4)$, namely if $U_B^T U_B = I_4$, then U belongs to K and there is no need for Cartan decomposition. Therefore we assume $U_B \notin \mathrm{SO}(4)$. Then observe that

$$U_B = Q^\dagger U Q = Q^\dagger k_2 Q \cdot Q^\dagger h Q \cdot Q^\dagger k_1 Q = O_2 h_D O_1, \tag{120}$$

where $O_i \equiv Q^\dagger k_i Q \in \mathrm{SO}(4)$ and $h_D \equiv Q^\dagger h Q$ is a diagonal matrix. From

$$U_B^T U_B = O_1^T h_D^2 O_1, \tag{121}$$

we notice that $U_B^T U_B$ is diagonalized by the orthogonal matrix O_1 and its eigenvalues constitute the diagonal elements of h_D^2. It implies that O_1 and h_D^2 can be determined from the eigenvalues and eigenvectors of $U_B^T U_B$. Note, however, that there is some ambiguities in the definition of h_D when taking square root of h_D^2. It can be shown[28] that there always exists a solution in which $t_x \geq t_y \geq |t_z|$, namely, the point (t_x, t_y, t_z) is assumed to be in the Weyl chamber of $\mathfrak{su}(4)$.[29] Then $k_1 = QO_1Q^\dagger$ and $h = Qh_DQ^\dagger$ are found straightforwardly. Finally, O_2 is fixed as $O_2 = U_B(h_DO_1)^{-1}$ and $k_2 = QO_2Q^\dagger$ is obtained.

5.3.1. *Examples*

As a first example, we consider a trivial case of the Walsh-Hadamard gate H_2 for two qubits given in (Eq. 68). We need to know first if H_2 is implemented with or without J-coupling operations. The solution is easily found by calculating

$$Q^\dagger H_2 Q = \begin{pmatrix} 1 & 0 & 0 & 0 \\ 0 & 0 & 0 & 1 \\ 0 & 0 & -1 & 0 \\ 0 & 1 & 0 & 0 \end{pmatrix} \tag{122}$$

and

$$U_B^T U_B = \left(Q^\dagger H_2 Q\right)^T \left(Q^\dagger H_2 Q\right) = I_4. \tag{123}$$

It implies that $H_2 \in \mathrm{SO}(4)$ and is implemented without any J-coupling operations as we have already shown in Eq. (69).

As a second example, let us determine NMR pulse sequence for the controlled-phase gate

$$U_{\mathrm{CP}} = \begin{pmatrix} 1 & 0 & 0 & 0 \\ 0 & 1 & 0 & 0 \\ 0 & 0 & 1 & 0 \\ 0 & 0 & 0 & -1 \end{pmatrix}. \tag{124}$$

The controlled-phase gate is employed to implement a CNOT gate along with one-qubit operations.[1] We find

$$U_B^T U_B = \left(Q^\dagger U_{\mathrm{CP}} Q\right)^T \left(Q^\dagger U_{\mathrm{CP}} Q\right) = \begin{pmatrix} -1 & 0 & 0 & 0 \\ 0 & 1 & 0 & 0 \\ 0 & 0 & 1 & 0 \\ 0 & 0 & 0 & -1 \end{pmatrix} \neq I_4. \tag{125}$$

Therefore, the NMR pulse sequence to implement U_{CP} involves two-qubit operations, such as U_{xx}, U_{yy}, or U_{zz}. The eigenvalues of $U_B^T U_B$ are $1, -1, -1, 1$ and the corresponding eigenvectors are $(0,1,0,0)^T$, $(1,0,0,0)^T$, $(0,0,0,1)^T$, $(0,0,1,0)^T$. Therefore,

$$h_D^2 = \begin{pmatrix} 1 & 0 & 0 & 0 \\ 0 & -1 & 0 & 0 \\ 0 & 0 & -1 & 0 \\ 0 & 0 & 0 & 1 \end{pmatrix}, \tag{126}$$

and

$$O_1 = \begin{pmatrix} 0 & 1 & 0 & 0 \\ 1 & 0 & 0 & 0 \\ 0 & 0 & 0 & 1 \\ 0 & 0 & 1 & 0 \end{pmatrix}, \tag{127}$$

are obtained. Note that this is one of the possible choices. We choose it since h_D takes a form

$$h_D = \begin{pmatrix} 1 & 0 & 0 & 0 \\ 0 & i & 0 & 0 \\ 0 & 0 & i & 0 \\ 0 & 0 & 0 & 1 \end{pmatrix}, \tag{128}$$

so that we can construct h_D as

$$h_D = e^{i\pi/4} U_{zz}(\pi/J) \tag{129}$$

using Table 5. Then, O_2 should be

$$O_2 = U_B (h_D O_1)^{-1} = \begin{pmatrix} 0 & 0 & 1 & 0 \\ 1 & 0 & 0 & 0 \\ 0 & 0 & 0 & 1 \\ 0 & -1 & 0 & 0 \end{pmatrix}, \tag{130}$$

which is certainly an element of SO(4). The element k_1 is easily obtained as

$$k_1 = Q O_1 Q^\dagger = \begin{pmatrix} 0 & 0 & -i & 0 \\ 0 & 0 & 0 & i \\ i & 0 & 0 & 0 \\ 0 & -i & 0 & 0 \end{pmatrix} = Y_1 Y_1 \bar{Z}_2 \bar{Z}_2. \tag{131}$$

Similarly we obtain

$$k_2 = \begin{pmatrix} 0 & 0 & -i & 0 \\ 0 & 0 & 0 & i \\ i & 0 & 0 & 0 \\ 0 & -i & 0 & 0 \end{pmatrix} = Z_1 Y_1 Y_1 \bar{Z}_2, \tag{132}$$

where we used the notations defined in Eqs. (66) and (67). Combining the above calculations, we obtain the NMR pulse sequence to implement $U_{\rm CP}$ as

$$\begin{aligned} k_2\, h\, k_1 &= \left(Z_1 Y_1 Y_1 \bar{Z}_2\right) \left(e^{i\pi/4} U_{zz}(\pi/J)\right) \left(Y_1 Y_1 \bar{Z}_2 \bar{Z}_2\right) \\ &= -e^{i\pi/4} Z_1 Z_2 Y_1 Y_1 U_{\rm E}(\pi) Y_1 Y_1, \end{aligned} \tag{133}$$

where we employed the rules given in Eq. (33) to simply the pulse sequence.

5.4. *Grover's database search algorithm*

5.4.1. *Background*

Suppose there is an unstructured database of N entries. One of the entries satisfies a given condition but we do not know which one is the desired entry. The database search is a task to find the entry that satisfies the condition. Classically, the only possible method to find the entry from the unsorted database is

(1) Pick up any entry in the database.
(2) Check whether it satisfies the condition.
(3) If it satisfies the condition, it is the entry we are looking for. If not, repeat (1) and (2) till the desired entry is found.

The condition is defined by a function $f(i)$, which is called an oracle, such that $f(i) = 0$ or 1 for each entry i and $f(w) = 1$ is realized only for one desired entry w. The target entry w is called a "file" to be found. In the worst case $N-1$ times queries (the steps (1) through (2)) are necessary and the average number of necessary queries is $N/2$. In 1996 Grover[30] proposed a quantum algorithm for the database search, with which one can find the desired entry with as small as $\sim \sqrt{N}$ queries.

In quantum information processing the set of N data entries is represented by an N-dimensional space spanned by orthogonal basis representing each entry. Then, a database search problem is reduced to a task to find the state $|w\rangle$ that represents the entry satisfying the condition $f(w) = 1$.

Grover's database search algorithm[30] consists of the following procedures:

(1) Prepare the initial state $|s\rangle$ which is a superposition of all the entries $|i\rangle$:

$$|s\rangle = \frac{1}{\sqrt{N}} \sum_i |i\rangle . \tag{134}$$

If $N = 2^m$, then $|s\rangle$ can be easily prepared by the Walsh-Hadamard gate as

$$|s\rangle = H^{\otimes m} |0\rangle^{\otimes m} . \tag{135}$$

(2) Let a unitary operator

$$
\begin{aligned}
G &= \Big(2\,|s\rangle\,\langle s| - I\Big)\left(\sum_i (-1)^{f(i)}\,|i\rangle\,\langle i|\right) \\
&= \Big(2\,|s\rangle\,\langle s| - I\Big)\Big(I - 2\,|w\rangle\,\langle w|\Big)
\end{aligned}
\tag{136}
$$

act n ($\sim \sqrt{N}$) times on $|s\rangle$. Here I is the identity operator. The action of G on $|\phi\rangle = \sum_i c_i|i\rangle$ with $\sum_i |c_i|^2 = 1$ enhances the coefficient c_w while suppressing the other $c_i(i \neq w)$, assuming all c_i are positive, as

$$
\begin{aligned}
G|\phi\rangle &= \Big(2\,|s\rangle\,\langle s| - I\Big)\Big(I - 2\,|w\rangle\,\langle w|\Big)|\phi\rangle \\
&= \Big(2\,|s\rangle\,\langle s| - I\Big)\Big(|\phi\rangle - 2c_w\,|w\rangle\Big) \\
&= 2\,|s\rangle\,\langle s\,|\phi\rangle - 4c_w\,|s\rangle\,\langle s\,|w\rangle - |\phi\rangle + 2c_w\,|w\rangle \\
&= 2\sum_{i,j,k}\frac{c_k}{N}\,|i\rangle\,\langle j\,|k\rangle - 4\frac{c_w}{N}\sum_{i,j}|i\rangle\,\langle j\,|w\rangle - \sum_i c_i\,|i\rangle + 2c_w\,|w\rangle \\
&= \sum_i (2\bar{c} - c_i - \frac{4c_w}{N})\,|i\rangle + 2c_w\,|w\rangle \\
&\simeq \sum_{i\neq w}(2\bar{c} - c_i)\,|i\rangle + (2\bar{c} + c_w)\,|w\rangle\,.
\end{aligned}
\tag{137}
$$

Here, $\bar{c} = (\sum_i c_i)/N$ or the mean value of c_i.

(3) After $n = [\pi/4\theta]$ iterations, where [ ] stands for the integer part and $\sin\theta = \sqrt{1/N}$, $\cos\theta = \sqrt{1 - 1/N}$, the probability amplitude localizes at $|w\rangle$, while all the other amplitude being negligible. Therefore the measurement of the state gives $|w\rangle$ with a probability very close to 1 and the database search is completed. Note that $\theta \sim 1/\sqrt{N}$ for a large enough N and the algorithm requires only $n \simeq \sqrt{N}$ steps.

Grover's algorithm can be understood more visually as follows.[31] Define a vector $|n\rangle$

$$
|n\rangle = \frac{1}{\sqrt{N-1}}\sum_{i\neq w}|i\rangle\,, \tag{138}
$$

which is orthogonal to $|w\rangle$. If we write the initial state $|s\rangle$ with $|n\rangle$ and $|w\rangle$,

$$
|s\rangle = \cos\theta\,|n\rangle + \sin\theta\,|w\rangle\,. \tag{139}
$$

Then, action of G gives

$$G(\alpha|n\rangle + \beta|w\rangle) = G\begin{pmatrix}\alpha\\ \beta\end{pmatrix}$$
$$= \begin{pmatrix}\cos 2\theta & -\sin 2\theta\\ \sin 2\theta & \cos 2\theta\end{pmatrix}\begin{pmatrix}\alpha\\ \beta\end{pmatrix}. \tag{140}$$

Therefore, Grover's iteration is regarded as repetition of rotations by an angle 2θ from $|s\rangle$ towards $|w\rangle$ in the 2-dimensional subspace spanned by $|n\rangle$ and $|w\rangle$. The number of necessary iterations is $n = \pi/4\theta \sim \pi\sqrt{N}/4$.

The algorithm for $N = 2^2 = 4$ entries is summarized in Table 6. The case of $N = 4$ is special in the sense that only one Grover's iteration gives the destination file.

Table 6. Quantum circuit for Grover's algorithm for $N = 2^2 = 4$ entries. R_{ij} is a selective inversion and U_{ij} is a total unitary operation representing the algorithm.

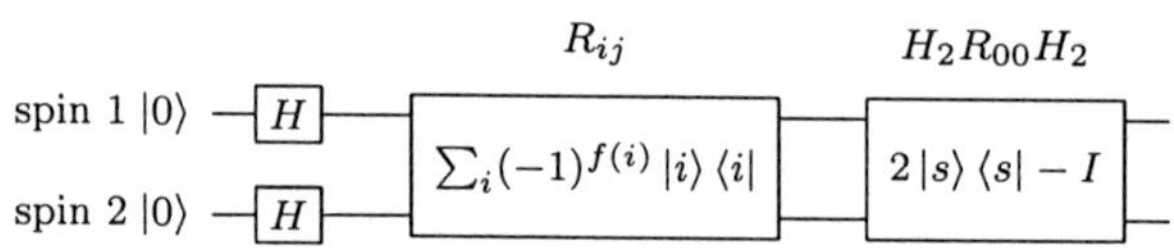

	00	01	10	11
R_{ij}	$\begin{pmatrix}1&0&0&0\\0&-1&0&0\\0&0&-1&0\\0&0&0&-1\end{pmatrix}$	$\begin{pmatrix}-1&0&0&0\\0&1&0&0\\0&0&-1&0\\0&0&0&-1\end{pmatrix}$	$\begin{pmatrix}-1&0&0&0\\0&-1&0&0\\0&0&1&0\\0&0&0&-1\end{pmatrix}$	$\begin{pmatrix}-1&0&0&0\\0&-1&0&0\\0&0&-1&0\\0&0&0&1\end{pmatrix}$
U_{ij}	$\begin{pmatrix}-1&0&0&0\\0&0&1&0\\0&1&0&0\\0&0&0&1\end{pmatrix}$	$\begin{pmatrix}0&0&1&0\\-1&0&0&0\\0&0&0&-1\\0&-1&0&0\end{pmatrix}$	$\begin{pmatrix}0&1&0&0\\0&0&0&-1\\-1&0&0&0\\0&0&-1&0\end{pmatrix}$	$\begin{pmatrix}0&0&0&1\\0&-1&0&0\\0&0&-1&0\\-1&0&0&0\end{pmatrix}$

5.4.2. *Implementation*

Chuang *et al.*[32] employed the pulse sequences shown in Table 7. Note that the matrices produced by their pulse sequences do not agree with those of actual Grover's algorithms: The relative phase of their matrix elements do not reproduce those in Table 6, although the difference does not affect NMR measurements.

We present our pulse sequences in Table 8, which are further simplified from the pulses of Chuang *et al* by employing the rules in Eq. (33). For

Table 7. Pulse sequences employed in Chuang *et al.* to implement Grover's algorithms. The pair (A,B) should be substituted by $(X,X),(\bar{X},X),(X,\bar{X})$, and $(\bar{X},\bar{X})$ for $w=00$, 01, 10, and 11, respectively.

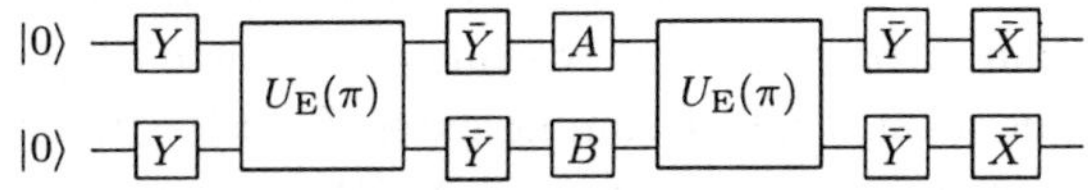

	00	01	10	11
U_{ij}	$\begin{pmatrix} -i & 0 & 0 & 0 \\ 0 & 0 & -i & 0 \\ 0 & -i & 0 & 0 \\ 0 & 0 & 0 & i \end{pmatrix}$	$\begin{pmatrix} 0 & 0 & 1 & 0 \\ 1 & 0 & 0 & 0 \\ 0 & 0 & 0 & 1 \\ 0 & -1 & 0 & 0 \end{pmatrix}$	$\begin{pmatrix} 0 & 1 & 0 & 0 \\ 0 & 0 & 0 & 1 \\ 1 & 0 & 0 & 0 \\ 0 & 0 & -1 & 0 \end{pmatrix}$	$\begin{pmatrix} 0 & 0 & 0 & i \\ 0 & i & 0 & 0 \\ 0 & 0 & i & 0 \\ -i & 0 & 0 & 0 \end{pmatrix}$

Table 8. Simplified pulse sequences for Grover's algorithms. The pair (A,B) should be substituted by $(\bar{X},Y),(\bar{X},\bar{Y}),(X,\bar{Y})$, and (X,Y) when $w=00$, 01, 10, and 11, respectively.

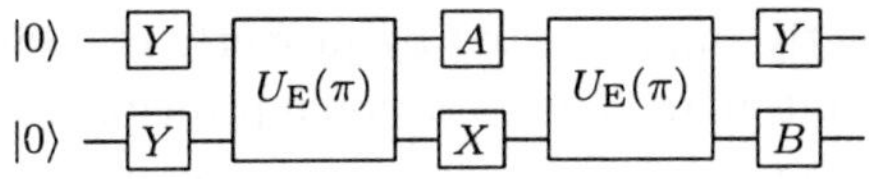

	00	01	10	11
U_{ij}	$\begin{pmatrix} -i & 0 & 0 & 0 \\ 0 & 0 & i & 0 \\ 0 & -i & 0 & 0 \\ 0 & 0 & 0 & -i \end{pmatrix}$	$\begin{pmatrix} 0 & 0 & i & 0 \\ i & 0 & 0 & 0 \\ 0 & 0 & 0 & -i \\ 0 & i & 0 & 0 \end{pmatrix}$	$\begin{pmatrix} 0 & i & 0 & 0 \\ 0 & 0 & 0 & -i \\ -i & 0 & 0 & 0 \\ 0 & 0 & -i & 0 \end{pmatrix}$	$\begin{pmatrix} 0 & 0 & 0 & i \\ 0 & i & 0 & 0 \\ 0 & 0 & i & 0 \\ -i & 0 & 0 & 0 \end{pmatrix}$

example, we simplify U_{01} as

$$\begin{aligned} U_{01} &= \bar{X}_1\bar{Y}_1\bar{X}_2\bar{Y}U_{\mathrm{E}}(\pi)\bar{X}_1\bar{Y}_1X_2\bar{Y}_2U_{\mathrm{E}}(\pi)Y_1Y_2 \\ &= \bar{X}_1\bar{Y}_1\bar{X}_2\bar{Y}U_{\mathrm{E}}(\pi)Z_1\bar{X}_1\bar{Z}_2X_2U_{\mathrm{E}}(\pi)Y_1Y_2 \\ &= \bar{Z}_2Z_1Y_1\bar{X}_1\bar{Y}_2X_2U_{\mathrm{E}}(\pi)\bar{X}_1X_2U_{\mathrm{E}}(\pi)Y_1Y_2 \\ &= \bar{Z}_2Z_2Z_1Z_1Y_1\bar{Y}_2U_{\mathrm{E}}(\pi)\bar{X}_1X_2U_{\mathrm{E}}(\pi)Y_1Y_2. \end{aligned} \tag{141}$$

The factors Z_i in the end of the sequence are irrelevant and can be removed to yield

$$U_{01} = Y_1\bar{Y}_2U_{\mathrm{E}}(\pi)\bar{X}_1X_2U_{\mathrm{E}}(\pi)Y_1Y_2. \tag{142}$$

Although we have reduced the number of pulses by the above simplification, the role played by each step in Grover's iteration becomes less clear then.

5.4.3. *Algorithm acceleration*

Here we briefly describe our contributions toward reduction of execution time in quantum computation. See the references [33,26] for further details.

As discussed in the previous section, we successfully reduced the number of pulses required to implement Grover's database search algorithm. Although the reduction in the number of pulses is quite helpful to reduce the gate operation errors, the total execution time is hardly changed since the number of the most time-consuming gates, $U_{\mathrm{E}}(\pi)$, is unchanged. Recall that J-coupling is the slowest process in NMR quantum computing. Here we show how to attain the optimal execution time and reduce the number of gates. Note that the shorter execution time is important to overcome decoherence.

The strategy, which we call "quantum algorithm acceleration" was originally proposed for fictitious Josephson charge qubits.[34,35] For a realization of a quantum algorithm U, we have to implement an n-qubit matrix which represents the quantum algorithm. Note that there is no necessity to implement individual elementary gates often employed to construct U. This matrix may be directly implemented by properly choosing the control parameters in the Hamiltonian. The variational principle tells us that the gate execution time can be reduced, in general, compared to the conventional construction with elementary gates since the conventional gate sequence is one of the possible solutions.

The quantum circuits for two-qubit Grover's algorithm is shown in the first row of Table 9. We obtained the optimal Cartan decomposition for Grover's algorithm by employing the method outlined in § 5.3. The result shows that the NMR execution times $T = \sum_{i=1}^{3} t_i$ form a discrete series as

$$T = (n+1)/J, \quad n = 0, 1, 2 \ldots \tag{143}$$

for all the "target files" $|00\rangle, |01\rangle, |10\rangle, |11\rangle$. The minimum execution time is $1/J$, which shows that the conventional pulse sequence in the first row of Table 7 is already time-optimal.

The initial state for Grover's algorithm is a pseudopure state prepared by applying cyclic permutations (CP) on a thermal state population as discussed in § 4.3.1. We merged the cyclic permutation gates U_{CP} and U_{CP^2} with Grover's search algorithm as shown in the second and third rows of the upper half of Table 9.

We obtained remarkable results when we optimized the pulse sequences for the combined unitary gates. The optimized quantum circuits are shown in the lower half of Table 9. Although we added extra gates to Grover's algorithm, the execution times became shorter than original Grover's search algorithms.

The experimental setup is the same as in the case of the Deutsch-Jozsa

Table 9. Pulse sequences of Grover's algorithm picking up the "file" $|01\rangle$. The initial state $|00\rangle$ is prepared as a pseudopure state by cyclic permutations of the state populations.

	Pulse sequence	Implemented matrix
Conventional		
U_{01}	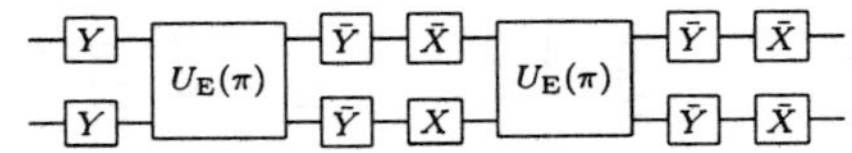	$\begin{pmatrix} 0 & 0 & 1 & 0 \\ 1 & 0 & 0 & 0 \\ 0 & 0 & 0 & 1 \\ 0 & -1 & 0 & 0 \end{pmatrix}$
$U_{01}U_{\mathrm{CP}}$	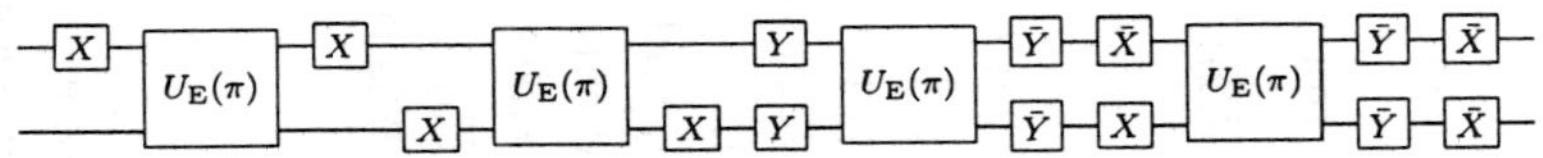	$\begin{pmatrix} 0 & -1 & 0 & 0 \\ -i & 0 & 0 & 0 \\ 0 & 0 & 1 & 0 \\ 0 & 0 & 0 & i \end{pmatrix}$
$U_{01}U_{\mathrm{CP}^2}$	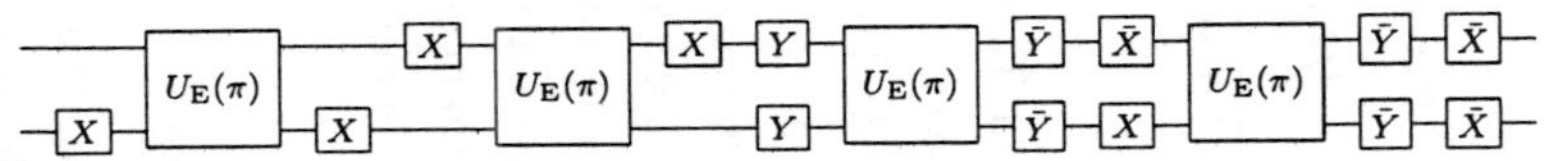	$\begin{pmatrix} 0 & 0 & 0 & -i \\ -i & 0 & 0 & 0 \\ 0 & 1 & 0 & 0 \\ 0 & 0 & 1 & 0 \end{pmatrix}$
Optimized		
U_{01}	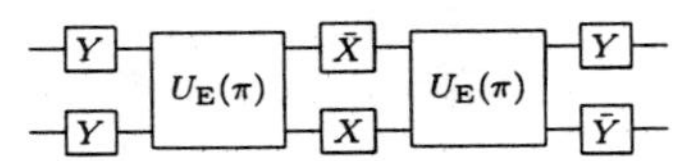	$\begin{pmatrix} 0 & 0 & i & 0 \\ i & 0 & 0 & 0 \\ 0 & 0 & 0 & -i \\ 0 & i & 0 & 0 \end{pmatrix}$
$U_{01}U_{\mathrm{CP}}$	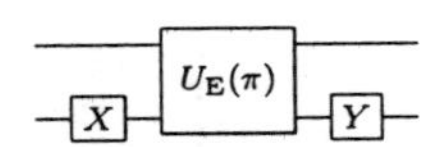	$\begin{pmatrix} 0 & -1 & 0 & 0 \\ -i & 0 & 0 & 0 \\ 0 & 0 & 1 & 0 \\ 0 & 0 & 0 & -i \end{pmatrix}$
$U_{01}U_{\mathrm{CP}^2}$	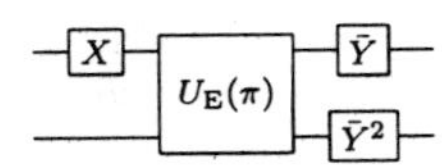	$\begin{pmatrix} 0 & 0 & 0 & -i \\ i & 0 & 0 & 0 \\ 0 & -1 & 0 & 0 \\ 0 & 0 & -1 & 0 \end{pmatrix}$

Note: The number of pulses in the optimized pulse sequences is reduced from 18 in our paper[33] to 11, here. Note, however, that the phases of the elements in the matrices realized with these pulse sequences do not coincide with those in Table 6 because z-rotations in the end of each sequence have been dropped as in Eq. (142).

algorithm. The sample molecule is cytosine solved in D_2O. The initial state is the thermal state. We performed the NMR pulse sequences listed in Table 9. The expected signals from the sample are given in Table 10.

Table 10. Expected signals which will be observed after performance of Grover's algorithm followed by the reading pulse on the spin 1 (Y_1). The sum of signals (temporal averaging) is taken to define the pseudopure state.

operation	signal
U_{01}	$\frac{1}{2}\cos(\omega_{0,1} - J/2)t + \frac{1}{2}\cos(\omega_{0,1} + J/2)t$
$U_{01}U_{\mathrm{CP}}$	$\cos(\omega_{0,1} + J/2)t$
$U_{01}U_{\mathrm{CP}^2}$	$-\frac{1}{2}\cos(\omega_{0,1} - J/2)t + \frac{1}{2}\cos(\omega_{0,1} + J/2)t$
Sum of the signals	$2\cos(\omega_{0,1} + J/2)t$

We applied pulse sequences listed in Table 9 on cytosine in the thermal state and observed the spectra shown in Fig. 10. Signals from spin 1 were measured and Fourier transformed. The left panel in Fig. 10 shows the results obtained with the conventional pulse sequences (the upper half of Table 9), while the right panel shows the results obtained with our accelerated algorithm (the lower half of Table 9). In both cases, we obtained signals that are in agreement with the theoretical expectation. This result proves that the quantum algorithm acceleration works correctly.

The sum of spectra of three performances, $\{U_{01}, U_{01}U_{\mathrm{CP}}, U_{01}U_{\mathrm{CP}^2}\}$ in Fig. 10, corresponds to that of $U_{01}|00\rangle$, where $|00\rangle$ is the pseudopure state created by temporal averaging.

5.4.4. *Warp-drive quantum computation*

We extended the idea of quantum algorithm acceleration further and actively reduced the execution time by adding an extra gate called the "warp-drive" gate.[26] When a permutation matrix W of the binary basis vectors is multiplied on the unitary gate U_{alg}, W may send U_{alg} to a point WU_{alg} near the identity matrix I so that it takes a shorter time to follow the time-optimal path connecting I with WU_{alg} than that connecting I with U_{alg} as illustrated in Fig. 11. We call this technique "warp-drive" since the reduction in the execution time takes place instantaneously as the extra gate is introduced. We have "warp-driven" U_{alg} to WU_{alg} by adding W to U_{alg}.

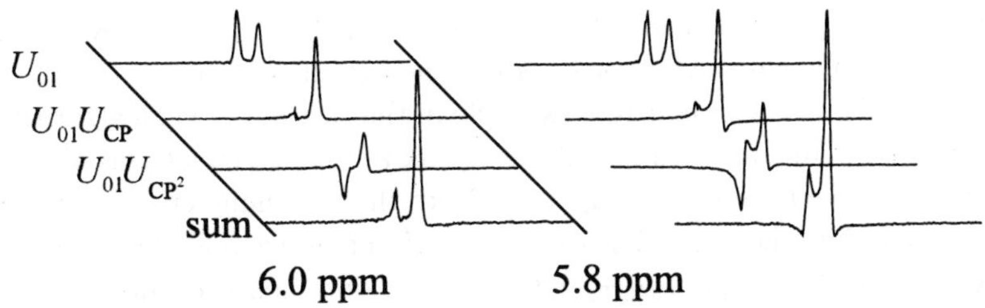

Fig. 10. Fourier transformed signals of the spin 1 after performance of Grover's database search algorithm. The left panel shows the results of the conventional pulse sequences, while the right panel are results of our accelerated pulse sequences. In both cases, the observed signals are in good agreement with the theoretically expected spectra listed in Table 10.

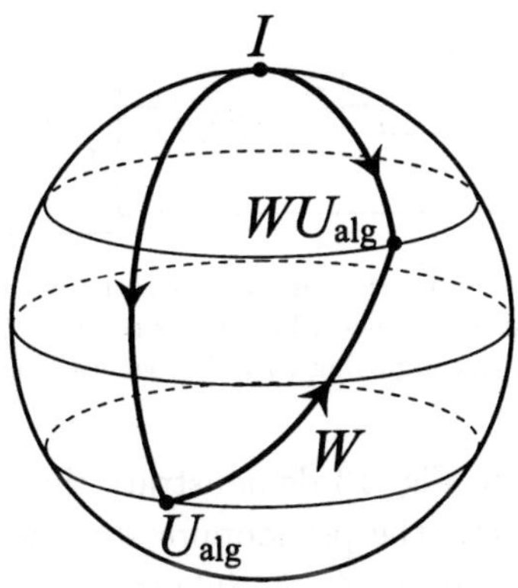

Fig. 11. A conceptual picture showing the action of the warp-drive gate W. The sphere represents the compact SU(4) group manifold. The warp-drive W sends a unitary matrix U_{alg} to WU_{alg}. The curves connecting these matrices represent time-optimal paths. The product unitary matrix WU_{alg} is reachable from the identity matrix I in a shorter execution time than U_{alg}.

The extra gate W which shortens the execution time must be simple enough so that we can deduce the output of $U_{\text{alg}}|0\rangle$ efficiently from the output of $WU_{\text{alg}}|0\rangle$ using classical computation only. Here $|0\rangle$ denotes the n-qubit fiducial state $|00\ldots0\rangle$. We call such gates warp-drive gates. Since a matrix $U_{\rm alg}$ is an element of a compact group $\mathrm{U}(2^n)$, there always exist such warp-drive gates which will reduce the execution time.

Let us show a more concrete example in the case of Grover's algorithms. When U_{10} is Cartan-decomposed, the optimized pulse sequence is obtained as shown in the first row in Fig. 5.4.4. Note that the number of $U_{\rm E}(\pi)$ gate is two and thus the execution time is $2\pi/J$. Then, we add the warp-drive gate ($U_{\rm CP} = U_{\rm CNOT12}\, U_{\rm CNOT21}$) to U_{10} and obtained the pulse sequence shown in the second row in Fig. 5.4.4. Note that the number of $U_{\rm E}(\pi)$ gate is one and thus the execution time is halved compared with that of U_{10}.

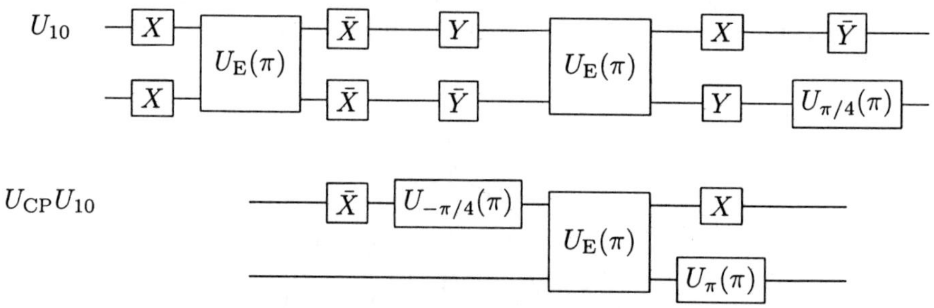

Fig. 12. Pulse sequences for Grover's algorithms without and with a warp-drive gate. The first row shows the pulse sequence which picks the file "10". The second row is the pulse sequence obtained by decomposing $U_{\rm CP}U_{10}$, where $U_{\rm CP}$ is the warp-drive gate.

The spectra shown in Fig. 13 demonstrate that the "warp-drive gate" works properly. This time, the pseudopure state was prepared by spatial labeling, which is discussed in § 4.3.2. Subsequently the gate U_{10} ($U_{\rm CP}U_{10}$) was applied on the pseudopure state $|00\rangle$. See the pulse sequence in the first (second) row in Fig. 5.4.4. The effectiveness of the "warp-drive" gate should be obvious from the fact that the signal of $U_{\rm CP}U_{10}$ is much sharper than that of U_{10}. Note that we should get $|11\rangle$ with $U_{\rm CP}U_{10}$ instead of $|10\rangle$.

6. DiVincenzo criteria

Here we briefly discuss whether the room-temperature liquid-state NMR quantum computation fulfills the DiVincenzo criteria[4]:

(1) *Be a scalable physical system with well-defined qubits:*

The NMR quantum computation employs the spins of atomic nuclei as qubits. We need to prepare appropriate molecules that contain spin-1/2 nuclei. Typical nuclei are hydrogen, fluorine, carbon-13, and nitrogen-15. The scalability, however, may

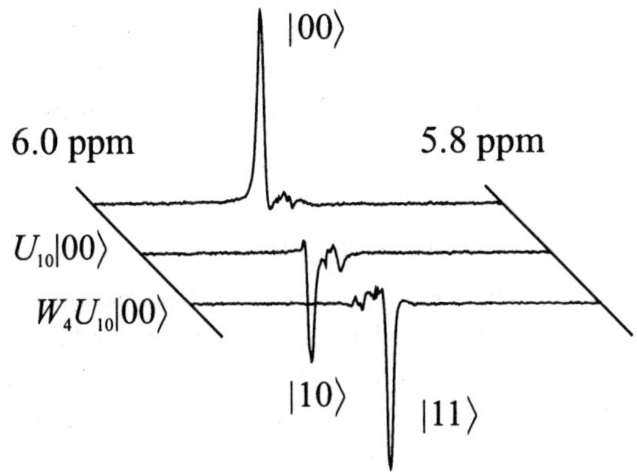

Fig. 13. Grover's database search algorithm with and without a warp-drive gate. The gates U_{10} and $U_{CP}U_{10}$ are applied on the pseudopure state $|00\rangle$, where U_{CP} is one of the warp-drive gates for U_{10}.

not be so obvious. The maximum number of qubits manipulated with a room-temperature liquid-state NMR quantum computer[2] so far is seven. The difficulty in scalability stems from the lack of an efficient method of initialization as is discussed next. It should be also noted that selective addressing to each qubit requires sufficient chemical shifts in resonance frequencies, which becomes more and more difficult as the number of spin increases.

(2) *Be initializable to a simple fiducial state such as* $|000\ldots\rangle$ *:*

Nuclear spins are subject to a strong magnetic field, typically on the order of 10 T, in thermal equilibrium at room temperature. The spins are in a highly *mixed state* far from a pure state required for quantum computations. One can prepare a pseudopure state, which effectively behaves like a pure state, by using temporal averaging, spatial averaging or logical labeling. These schemes, however, come with a cost; either the signal intensity decreases exponentially, or the number of required measurements grows exponentially as the number of qubits increases.

Another important issue related to a pseudopure state, or

equivalently a mixed-state, is the lack of entanglement.[3] Optical pumping and various solid-state NMR schemes may fulfill this criterion, but they are beyond the scope of the present discussion.

(3) *Have much longer decoherence times,* and

(4) *Have a universal set of quantum gates:*

The third and the fourth criteria are discussed together. One-qubit gates are realized with various rf pulses, whose duration is of the order of 10 μs, while the CNOT gate can be implemented with several rf pulses and time evolution by a J-coupling between two nuclei, which takes of the order of 10 ms in total. Therefore, the universal set of quantum gates is provided in NMR quantum computation.

Nuclear spins in a molecule dissolved in a liquid have very long coherence times of more than a few seconds, when properly prepared. The number of feasible gate operations within decoherence time may be reasonably large to execute simple algorithms.

(5) *Permit high quantum efficiency, qubit-specific measurements:*

Readout of nuclear spin states is a well established technique in NMR. It should be remembered, however, that readout is an ensemble measurement of many nuclear spins, which may require a modification of certain quantum algorithms.[36]

In summary, the room-temperature liquid-state NMR technique in its present form cannot be a true candidate for a scalable quantum computer unless a drastically new improvement is developed. However, its potential importance should not be underestimated. The difficulty arises from the fact that the spin polarization of molecules in a liquid at room temperature is very small, of the order of 10^{-5}. If one could have a highly polarized "molecules", NMR-inspired techniques would lead to a *true* quantum computer. These "molecules" are not necessarily natural but can be artificial. Electrons on surface of superfluid helium [37] can be a good candidate for such artificial molecules.

We regard an NMR quantum computer as a good simulator for realistic quantum computers. We use an NMR quantum computer with several qubits in our research to investigate algorithm acceleration, decoherence simulation, decoherence suppression and other important issues in the physical realization of quantum computers. We strongly believe that the role played by an NMR quantum computer is significant in the fundamental research in these fields.

Acknowledgements

We would like to thank Toshie Minematsu for assistance in NMR operations and Katsuo Asakura and Naoyuki Fujii of JEOL for assistance in NMR pulse programming. MN would like to thank partial supports of Grant-in-Aids for Scientific Research from the Ministry of Education, Culture, Sports, Science and Technology, Japan, Grant No. 13135215 and from Japan Society for the Promotion of Science (JSPS), Grant No. 14540346. ST is partially supported by JSPS, Grant Nos. 15540277 and 17540372.

References

1. M. A. Nielsen, and I. L. Chuang, *Quantum Computation and Quantum Information*, (Cambridge University Press, Cambridge, 2000).
2. L. M. K. Vandersypen, M. Steffen, G. Breyta, C. S. Yannonl, M. H. Sherwood, and I. L. Chuang, Nature **414**, 883 (2001).
3. S. L. Braunstein, C. M. Caves, R. Jozsa, N. Linden, S. Popescu, and R. Schack, Phys. Rev. Lett. **83**, 1054 (1999).
4. D. P. DiVincenzo, http://www.research.ibm.com/ss_computing/. See, also, D. P. DiVincenzo, Science **270**, 255 (1995).
5. D. G. Cory, A. F. Fahmy, and T. F. Havel, Proc. Natl. Acad. Sci. USA **94**, 1634, March (1997).
6. J. A. Jones, Prog. NMR Spectrosc. **38**, 325 (2001).
7. L. M. K. Vandersypen and I. L. Chuang, Rev. Mod. Phys. **76**, 1037 (2004).
8. L. M. K. Vandersypen, Stanford University Thesis (2001).
9. For example, see T. E. W. Claridge, *High-Resolution NMR techniques in Organic Chemistry*, (Elsevier, Amsterdam, 2004)
10. R. R. Ernst, G. Bodenhausen, and A. Wokaun, *Principles of Nuclear Magnetic Resonance in One and Two Dimensions*, (Oxford University Press, Oxford, 1991).
11. F. Bloch and A. Siegert, Phys. Rev. **57**, 522 (1940).
12. N. F. Ramsey, Phys. Rev. **100**, 1191 (1955).
13. R. Laflamme, E. Knill, D. G. Cory, E. M. Fortunato, T. F. Havel, C. Miquel, R. Martinez, C. J. Negrevergne, G. Ortiz, M. A. Pravia, Y. Sharf, S. Sinha, R. Somma, and L. Viola, Los Alamos Science Number **27**, 226 (2002).
14. D. G. Cory, R. Laflamme, E. Knill, L. Viola, T. F. Havel, N. Boulant, G. Boutis, E. Fortunato, S. Lloyd, R. Martinez, C. Negrevergne, M. Pravia, Y. Sharf, G. Teklemariam, Y. S. Weinstein, and W. H. Zurek, Fortschr. Phys. **40**, 875 (2000), J. A. Jones, Fortschr. Phys. **40**, 909 (2000).
15. H. De Raedt, K. Michielsen, A. Hams, O. Miyashita, and K. Saito, Eur. Phys. J. B**27**, 15 (2002).

16. A. Barenco, C.H. Bennett, R. Cleve, D.P. DiVincenzo, N. Margolus, P. Shor, T. Sleator, J. Smolin and H. Weinfurter, Phys. Rev. A **52**, 3457 (1995).
17. M. A. Pravia, E. Fortunato, Y. Weinstein, M. D. Price, G. Teklemariam, R. J. Nelson, Y. Sharf, S. Somaroo, C. H. Tseng, T. F. Havel, D. G. Cory, Concepts Magn. Res. **11**, 225 (1999).
18. U. Sakaguchi, H. Ozawa, and T. Fukumi, Phys. Rev. A **61**, 042313 (2000).
19. Y. Sharf, T. F. Havel, D. G. Cory, Phys. Rev. A **62**, 052314 (2000).
20. N. A. Gershenfeld and I. L. Chuang, Science **275**, 350 (1997), L. M. K. Vandersypen, C. S. Yannoni, M. H. Sherwood, I. L. Chuang, Phys. Rev. Lett. **83** (1999) 3085.
21. D. Deutsch and R. Jozsa, Proc. Roy. Soc. London, Ser. A bf 439 553 (1992).
22. http://www.jeol.com/nmr/nmr.html.
23. J. A. Jones, M. Mosca, and R. H. Hansen, J. Chem. Phys. **109**, 1648 (1998). J. A. Jones and M. Mosca, Nature **393**, 344 (1998).
24. I. L. Chuang, L. M. K. Vandersypen, X. Zhou, D. W. Leung, and S. Lloyd, Nature **393**, 143 (1998).
25. N. Khaneja, R. Brockett, and S. J. Glaser, Phys. Rev. A **63**, 032308 (2001).
26. M. Nakahara, J. J. Vartiainen, Y. Kondo, S. Tanimura, and K. Hata, Phys. Lett. A, to be published and eprint quant-ph/0411153.
27. Y. Makhlin, Quant. Info. Proc. **1**, 243 (2002).
28. A. M. Childs, H. L. Haselgrove, and M. A. Nielsen, Phys. Rev. A **68**, 052311 (2003).
29. J. Zhang, J. Vala, S. Sastry, and K. B. Whaley, Phys. Rev. A **67**, 042313 (2003).
30. L. K. Grover, Phys. Rev. Lett. **79**, 325 (1997).
31. E. Farhi and S. Gutmann, Phys. Rev. A **57**, 2403 (1998).
32. I. L. Chuang, N. Gershenfeld, M. Kubnec, Phys. Rev. Lett. **80**, 3408 (1998).
33. M. Nakahara, Y. Kondo, K. Hata, and S. Tanimura, Phys. Rev. A **70**, 052319 (2004).
34. A. O. Niskanen, J. J. Vartiainen, and M. M. Salomaa, Phys. Rev. Lett. **90**, 197901 (2003).
35. J. J. Vartiainen, A. O. Niskanen, M. Nakahara, and M. M. Salomaa, Phys. Rev. A **70**, 012319 (2004).
36. N. A. Gershenfeld and I. L. Chuang, Science **275**, 350 (1997).
37. P. M. Platzman, M. I. Dykman, Science **284** 1967(1999).

Optical Quantum Computation

Kae Nemoto

National Institute of Informatics,
2-1-2 Hitotsubashi, Chiyoda-ku, Tokyo 101-8430, Japan
E-mail: nemoto@nii.ac.jp

W. J. Munro

Hewlett-Packard Laboratories,
Filton Road, Stoke Gifford, Bristol BS34 8QZ, UK
E-mail: bill.munro@hp.com

T. P. Spiller

Hewlett-Packard Laboratories,
Filton Road, Stoke Gifford, Bristol BS34 8QZ, UK
E-mail: timothy.spiller@hp.com

We review recent theoretical progress in finding ways to perform quantum processing with optics using photon encoded qubits, qudits and continuous variables. We discuss the requirements, advantages and disadvantages of each approach and show that optics holds significant promise for computation.

1. Introduction

There are currently numerous possible routes forward for quantum computation and communication hardware[1,2]. A significant number of these are based on coherent condensed matter systems[3], however another strong candidate (especially for communication) would be to use states of the electromagnetic field (photons). Optics has already played a major role in the testing of fundamental properties of quantum mechanics[5] and, more recently, implementing simple quantum information protocols. This has been possible because photons are easily produced and manipulated and, as the electro-magnetic environment at optical frequencies can be regarded as vacuum, are relatively decoherence free. If we consider quantum information processing and computation using photons then we have a number of distinct techniques we can use to encode and process this information. These include but are not limited to:

- single photon linear optics
- single photon nonlinear optics
- qudit information processing
- continuous variable (CV) information processing (qunat information processing)
- Hybrid systems using qubits, qudits and qunats.

Now that we have started to consider quantum information processing with light, we need to define a set of criteria to discuss to help us evaluate its potential. Essentially all aspects of quantum information processing can be reduced to three stages: prepare; evolve; measure. As we are classical objects, and can only relate to conventional information, we have to carefully prepare our quantum information (qubits, qudits or qunats) in appropriate states, allow them to evolve (feeding in the classical information pertaining to the process), and then measure them to extract information in a form we can use. A well known set of requirements that need to be satisfied so this information processing can be performed has been summarized in the universally accepted DiVincenzo Criteria[6,7]:

(1) A collection of well-characterized qubits, qudits or qunats is needed. One at a time will do for cryptography although entangled pairs could be a useful luxury. Controlled interactions between a few qubits, qudits, qunats are needed for small scale processing. Scalability of the number of qubits, qudits or qunats is necessary for full blown quantum computing.
(2) Preparation of known initial states for the qubits, qudits or qunats must be possible. The purer the better.
(3) The quantum coherence of the system(s) must be maintained to a high degree during the evolution/computation stage, giving a high fidelity for the final state. For few-qubit processing it may suffice to have a straight shot at the process with good qubits, qudits or qunats and gates; for large scale quantum computation error correction will almost certainly be needed[8]. For fault-tolerant operation the fidelity of individual gates probably needs to be 0.999 or better.
(4) The unitary quantum evolution required by the algorithm or protocol must be realizable. Some unitary processes are much easier in experiment than others. As with conventional computing, the minimum of a universal set of elementary gates must be possible.

(5) High fidelity quantum measurements on specific qubits, qudits or qunats must be possible, in order to read out the result of the process or computation.
(6) The capability to interconvert stationary (processing or memory) and flying (communication) quantum information must exist.
(7) It must be possible to transmit flying qubits, qudits or qunats coherently between specified locations.

For a localized quantum processing and computing device the first five criteria (1)-(5) needed to be satisfied, with the last two (6) and(7) required for genuinely spatially separated processing. Criterion (4) indicates that *the unitary quantum evolution required by the algorithm or protocol must be realizable.* However there may be certain algorithms that do not need a universal set of quantum gates to implement them. The potential problem with this is that without the use of a universal set of quantum gates, the quantum process may be able to be efficiently classically simulated. Just because our quantum processes create entangled states does not mean that this process can not be efficiently classically simulated. The concept of what can be efficiently classically simulated is summarized in the Gottesman-Knill (GK) theorem[9,10] which we will discuss below.

1.1. *The Gottesman-Knill theorem*

It is clear that certain quantum information processing tasks can be implemented much more easily than others. Simple schemes in quantum communication require only superposition for the quantum schemes to be distinguished from their corresponding classical ones. However, for computational tasks, simply having superpositions of quantum states is not enough to achieve truly quantum computation, distinguished from classical computation. We need to consider the concept of universality.

Universality is an important concept that allows quantum computational circuits to be distinguished from their classical counterparts. The GK theorem[9,10] states that any quantum algorithm that initiates in the computational basis and employs only a restricted class of gates (Hadamard, phase, CNOT, and Pauli gates), along with projective measurements in the computational basis, can be efficiently simulated on a classical computer. This means there is no computational advantage in using such circuits on a quantum computer. A classical machine could simulate the circuit efficiently. This set of gates described above preserves the Pauli group, which consists of Pauli operators and the identity operator I with coefficients of $\{\pm 1, \pm i\}$. For instance, the Pauli group for one qubit is $\{\pm\sigma_x, \pm\sigma_y, \pm\sigma_z, \pm I, \pm i\sigma_x, \pm i\sigma_y, \pm i\sigma_z, \pm iI\}$. Hence the GK theorem in discrete variable (and more especially qubit) quantum information provides a valuable tool for assessing the classical complexity of a given process. The set of gates in the GK theorem obviously does not satisfy the universality requirements, however addition of a simple single-qubit $\pi/8$ gate can make the set universal.

Optics has geometrical and structural freedom to design quantum computational circuits. This is a large advantage in optics. In particular, optical implementation allows us to use qubits, qudits and qunats freely. It is known that qudit computation is very analogous to qubit computation, and usually these are interchangeable with each other, while qunat computation can be quite distinct from discrete variable quantum computation.

Following the discussion of qubit computation, we can generalize the GK theorem to qunat (continuous variable) computation[11]. To generalize the theorem, we first define the generalized Pauli group. Unlike the discrete Pauli group for qubits, the generalized Pauli group is a continuous (Lie) group, and can only be generated by a set of continuously-parameterized operators. The generators of this group are the $2n$ canonical operators $\hat{q}_i$, $\hat{p}_i$, $i = 1, \ldots, n$, along with the identity operator $\hat{I}$, satisfying the commutation relations $[\hat{q}_i, \hat{p}_j] = i\hbar\delta_{ij}\hat{I}$. Here the subscripts indicate different modes of signal light in optical implementation. For a single oscillator, we have the canonical operators $\{\hat{q}, \hat{p}, \hat{I}\}$ which generate the single oscillator Pauli operators

$$X(q) = e^{-\frac{i}{\hbar}q\hat{p}}, \quad Z(p) = e^{\frac{i}{\hbar}p\hat{q}}, \tag{1}$$

with $q, p \in \mathbb{R}$. Based on the generalized Pauli operators, we can now find operations which preserve the generalized Pauli operators. This can be done by generalizing the gate set of the GK theorem for qubits. The Fourier transform F is the qunat analog of the Hadamard transformation. It is defined as

$$F = \exp\left[\frac{i}{\hbar}\frac{\pi}{4}(\hat{q}^2 + \hat{p}^2)\right], \tag{2}$$

and the action on the Pauli operators is

$$\begin{aligned} FX(q)F^\dagger &\rightarrow Z(q) \\ FZ(p)F^\dagger &\rightarrow X(p)^{-1}. \end{aligned} \tag{3}$$

The "phase gate" $P(\eta)$ for qunats is a squeezing operation, defined by

$$P(\eta) = \exp\left[\frac{i}{2\hbar}\eta\hat{q}^2\right], \tag{4}$$

and the action on the Pauli operators is

$$
\begin{aligned}
P(\eta)X(q)P(\eta)^{\dagger} &\rightarrow e^{\frac{i}{2\hbar}\eta q^{2}}X(q)Z(\eta q)\,,\\
P(\eta)Z(p)P(\eta)^{\dagger} &\rightarrow Z(p)\,.
\end{aligned}
\tag{5}
$$

(The operator $P(\eta)$ is called the phase gate, in analogy to the discrete-variable phase gate P, because of its similar action on the Pauli operators.)

The two-bit operation, CNOT gate, can be generalized to the SUM gate as

$$
\mathrm{SUM} = \exp\Bigl(-\frac{i}{\hbar}\hat{q}_1 \otimes \hat{p}_2\Bigr)\,.
\tag{6}
$$

The action of this gate on the Pauli group for two systems is given by

$$
\begin{aligned}
\mathrm{SUM}X_1(q)\otimes I_2\mathrm{SUM}^{\dagger} &\rightarrow X_1(q)\otimes X_2(q)\,,\\
\mathrm{SUM}Z_1(p)\otimes I_2\mathrm{SUM}^{\dagger} &\rightarrow Z_1(p)\otimes I_2\,,\\
\mathrm{SUM}I_1\otimes X_2(q)\mathrm{SUM}^{\dagger} &\rightarrow I_1\otimes X_2(q)\,,\\
\mathrm{SUM}I_1\otimes Z_2(p)\mathrm{SUM}^{\dagger} &\rightarrow Z_1(p)^{-1}\otimes Z_2(p)\,.
\end{aligned}
\tag{7}
$$

Now, the generalized GK theorem for qunats can be summarized as follows: Any qunat quantum information process that begins with Gaussian states (products of squeezed displaced vacuum states) and performs only

- linear phase-space displacements (given by the Pauli group),
- squeezing transformations on a single oscillator system,
- SUM gates,
- measurements in the position- or momentum-eigenstate basis (measurements of Pauli group operators) with finite errors and losses,
- classical feedforward,

can be efficiently simulated using a classical computer. The conditions can be simplified by transformations generated by inhomogeneous quadratic Hamiltonians in terms of the canonical operators $\{\hat{q}_i, \hat{p}_i;\, i = 1,\ldots,n\}$. Thus, any circuit built up of components described by one- or two-mode quadratic Hamiltonians [such as the set of gates SUM, F, $P(\eta)$, and $X(q)$], that begins with finitely squeezed states and involves only measurements of canonical variables may be efficiently classically simulated.

1.2. *Universal Quantum Computation*

Implementing a universal set of gates is certainly a way to achieve universal quantum computation. However, in real quantum mechanical systems it is generally difficult to implement such a universal gate set. Generally speaking, systems with small decoherence have difficulties with controlled operations between two (or more) qubits, and where such controlled operation are natural, it is often difficult to maintain quantum coherence in the system and to ensure the accessibility to each individual qubit. Given the fact that universal quantum computation requires both good quantum coherence and entanglement in the computational system, this trade-off relation between quantum coherence and multi-qubit interaction seems to be an obstacle to universal quantum computation, when one considers the requirement for universal gate sets. However, implementation of universal gate sets is not the only way to achieve universal quantum computation. In fact, some implementations of non-universal gate sets can achieve universal quantum computation in a scalable manner. Such implementations are not allowed to access to the entire Hilbert space, yet it is possible for these gates to simulate universal quantum computation on a lower dimensional subspace. For qubits restricted to states with just real amplitudes, which is called a rebit subspace, the controlled-rotation gate is such an example[12].

Universal quantum computation can be achieved through various approaches. In addition to gate-based schemes, universal quantum computation can also be achieved by measurement alone, or through schemes based on measurements[13–15]. In classical computation, measurement schemes are trivial, and hence universal classical computation is considered just in terms of gate-based schemes. However, in quantum computation this is not the case. Measurement alone can be as powerful as a universal gate set. A number of measurement-based universal quantum computation schemes have been proposed. Universal quantum computation can be broadly classified in three categories[16]:

- Universal set: A universal set can construct an arbitrary circuit with an arbitrary precision.
- Universal computational set: A universal computational set is not universal by the definition of universality, but can simulate universal computation with polynomial extra resource.
- Non-universal set: A non-universal set can be efficiently simulated on a classical computer and cannot simulate universal computation with polynomial extra resource. It can construct universal computation only with additional measurement strategies.

Now because of the power of measurement in quantum computation, it is obvious that it does not make much sense to evaluate quantum computation schemes solely either on gate sets or on measurement schemes. This aspect of quantum computation can be exploited to construct a circuit bypassing the difficulties of some particular operations. However, at the same time it adds an extra complication into our criteria for universal quantum computation.

1.3. *Back to the DiVincenzo criteria*

Now let us return to our discussion of the DiVincenzo criteria. The demands from these seven criteria are extremely tough indeed and the construction of small but useful quantum processors will certainly not be easy. Nevertheless, the goal of performing quantum simulations beyond what we can do with any conventional computer seems to be a good challenge for the next decade of research. If we can get that far, it will be rather easier to make predictions about the bigger goal of large scale many-qubit computing.

This article will be structured into four main sections, the first being on linear optical quantum computation, the second being on continuous variable quantum computation and the third on hybrid schemes that combine both single photons and continuous variables. This article will finish with a brief discussion of recent experiments and the prospects for optical quantum information processing.

2. Single-photon quantum computation: processing using only linear optics

Now let us begin our discussion of quantum computing with single photons. As we mentioned in our introduction, optics with single photons seems to be a natural candidate for quantum computation and information processing. For decades, the photon has been studied extensively by theorists and experimentalists alike and its properties are well known. Polarization information can be encoded on a single photon easily and this leads to an excellent qubit. For a polarization-encoded qubit an arbitrary state can be represented as

$$\cos\theta|H\rangle + e^{i\phi}\sin\theta|V\rangle \tag{8}$$

where $|H\rangle$ and $|V\rangle$ represent the horizontal and vertical polarization of that single photon and are the basis states for the qubit. The general state is parametrized by two angles θ and ϕ. This space can be pictured as the surface of a sphere, as illustrated in Figure 1. One of the greatest advantages of encoding in polarization is that single qubit operations are deterministic and trivial. They are simply achieved with linear elements such as

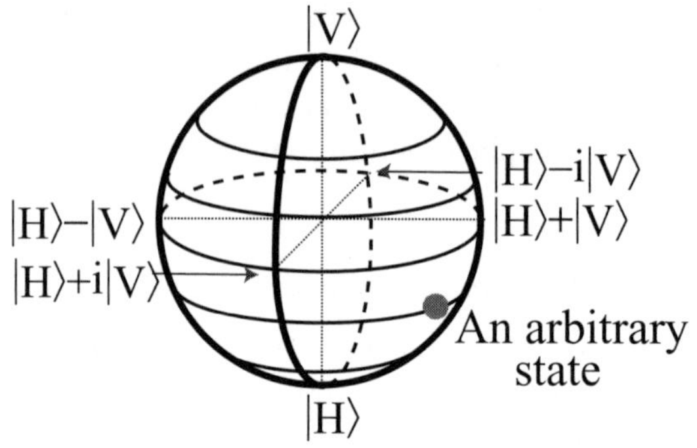

Fig. 1. The state space of a polarisation-encoded qubit. Classical bits only have access to the two polar states, but an arbitrary qubit state can be anywhere on the surface of the sphere. Examples of particular (unnormalized) equatorial states are labeled.

beam-splitters, polarizing beam-splitters and phase shifters. Unfortunately, an entangling two-photon gate such as the CNOT gate is far less straightforward, because, in ordinary materials, photons do not interact with each other.

The lack of nonlinearity had been a fatal problem in the optical implementation of a universal quantum computer until a recent proposal by Knill, Laflamme and Milburn[17]. This linear optics quantum computation scheme exploits a new non-deterministic process to generate nonlinearity only from linear elements, single-photon sources and photon number detection. These non-deterministic gates can then be teleported into the computational circuit allowing their operation to be deterministic in terms of the computation. We will now describe this protocol in detail.

2.1. *Linear optical quantum computation*

Recently, Knill, Laflamme and Milburn (KLM) [17] proposed an elegant scheme to sidestep the requirement for strong nonlinear materials. They showed that it is possible to generate an effective large nonlinear interaction between single photons using passive optical elements and photo-detection. In this scheme the optical mode of interest is combined with several other ancilla modes by passive linear optical circuits (beam-splitters) and conditioned on obtaining particular detector signatures in the output ancilla modes. The process is non-deterministic, but successful operation is flagged by the ancilla measurement result, that is to say, successful operation is heralded.

The basic operation they considered was a nonlinear sign shift on a general two-photon state, a state with two or less photons. This conditional operation transform the two-photon state according to

$$c_0|0\rangle + c_1|1\rangle + c_2|2\rangle \rightarrow c_0|0\rangle + c_1|1\rangle - c_2|2\rangle \tag{9}$$

where $|n\rangle$ is the nth Fock state of the optical field and the coefficients c_j are complex amplitudes which satisfy the usual normalization constraint $\sum |c_j|^2 = 1$. The implementation of this gate with linear optical elements and single photon detection is shown in Figure 2.

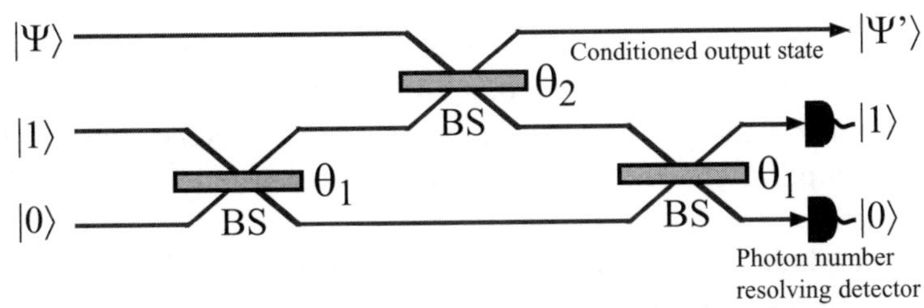

Fig. 2. Schematic diagram of the KLM nonlinear sign shift gate. This gate is composed of three input states, three beam-splitters and two photon number resolving detectors. The three input states are our unknown signal state $|\Psi\rangle = c_0|0\rangle + c_1|1\rangle + c_2|2\rangle$, plus two ancilla states, a single photon $|1\rangle$ and a vacuum state $|0\rangle$ respectively. These three states interact with each other via the beam-splitters characterized by θ_1, θ_2 and θ_1 (where $\cos^2\theta_1 = \frac{1}{4-2\sqrt{2}}$ and $\cos^2\theta_2 = 3 - 2\sqrt{2}$). After the interaction of the three states with the beam-splitters, the ancilla states are detected using single photon number resolving detectors. When a photon is detected in the first ancilla and the vacuum in the second ancilla, the signal state $|\Psi\rangle$ is transformed to $|\Psi'\rangle = c_0|0\rangle + c_1|1\rangle - c_2|2\rangle$. We observe the sign of the $|2\rangle$ amplitude has flipped and hence the name of the gate being a nonlinear sign shift.

Let us now consider the action of this gate in further detail. Consider that our signal state is the vacuum $|0\rangle$. After the interaction with the beam-splitters and the detection of one photon in the first ancilla and the vacuum in the second ancilla, the signal state is transformed according to

$$|0\rangle \rightarrow \left[\cos^2\theta_1 \cos\theta_2 + \sin^2\theta_1\right]|0\rangle \, . \tag{10}$$

Similarly the one-photon signal state $|1\rangle$ is transformed according to

$$|1\rangle \rightarrow -\left[\cos^2\theta_1 \cos 2\theta_2 + \sin^2\theta_1 \cos\theta_2\right]|1\rangle \, . \tag{11}$$

Lastly the two-photon signal state $|2\rangle$ is transformed according to

$$|2\rangle \rightarrow -\cos\theta_2 \left[\frac{1}{2}\cos^2\theta_1 \left\{3\cos 2\theta_2 - 1\right\} + \sin^2\theta_1 \cos\theta_2\right]|2\rangle \, . \tag{12}$$

Now if we choose $\cos^2\theta_1 = \frac{1}{4-2\sqrt{2}}$ and $\cos^2\theta_2 = 3 - 2\sqrt{2}$, then

$$|0\rangle \to \frac{1}{2}|0\rangle, \qquad |1\rangle \to \frac{1}{2}|1\rangle, \qquad |2\rangle \to -\frac{1}{2}|2\rangle \tag{13}$$

We see immediately that all three transformations have the same $\frac{1}{2}$ coefficient with the $|2\rangle$ component also having a negative sign. The $\frac{1}{2}$ squared is the probability that the transformation occurs. This means that the general state $|\Psi\rangle$ will be transformed according to

$$|\Psi\rangle = c_0|0\rangle + c_1|1\rangle + c_2|2\rangle \to \frac{1}{2}|\Psi'\rangle = \frac{1}{2}\left[c_0|0\rangle + c_1|1\rangle - c_2|2\rangle\right] . \tag{14}$$

The loss in amplitude reflects other outcomes, and so the transformation (9) is effected with a success probability of 1/4. This probabilistic but heralded nonlinear sign shift gate is the core gate in the KLM linear optical quantum computation scheme. From this gate the usual CNOT gate can be constructed.

Before we indicate how to construct the CNOT gate, which with single qubit gates forms a universal set of gates, we will examine several variants for the construction of nonlinear sign shift gates[18,19]. These gates are depicted in Figure 3. Both of these variant gates enable the sign to be flipped on the $|2\rangle$ amplitude but require one fewer beam-splitter. There is a slight cost to the efficiency of this variant gates due to having one less beam-splitter. The probability of success reduces from 25 percent to roughly 23 percent[18].

Now let's consider the construction of a CNOT gate from these nonlinear sign shift gates. The design is rather simple and is depicted in Figure 4. As a controlled-phase (CPhase) gate is equivalent to the CNOT gate up to local operations, we will consider this form first. The CPhase gate is composed of three basic elements, a 50/50 beam-splitter followed by two nonlinear sign shift gates and finally another 50/50 beam-splitter. Let us now examine how this gate acts on a general two qubit state (where our qubits have been encoded in photon number) $\beta_0|00\rangle + \beta_1|01\rangle + \beta_2|10\rangle + \beta_3|11\rangle$. The first qubit state represents the control qubit and the second the target. After the first 50/50 beam-splitter this state is converted (transformed) to

$$\beta_0|00\rangle + \frac{\beta_1}{\sqrt{2}}\left[|01\rangle - |10\rangle\right] + \frac{\beta_2}{\sqrt{2}}\left[|10\rangle + |01\rangle\right] + \frac{\beta_3}{\sqrt{2}}\left[|02\rangle - |20\rangle\right] . \tag{15}$$

Then applying separate nonlinear sign gates to both modes we get

$$\beta_0|00\rangle + \frac{\beta_1}{\sqrt{2}}\left[|01\rangle - |10\rangle\right] + \frac{\beta_2}{\sqrt{2}}\left[|10\rangle + |01\rangle\right] - \frac{\beta_3}{\sqrt{2}}\left[|02\rangle - |20\rangle\right] \tag{16}$$

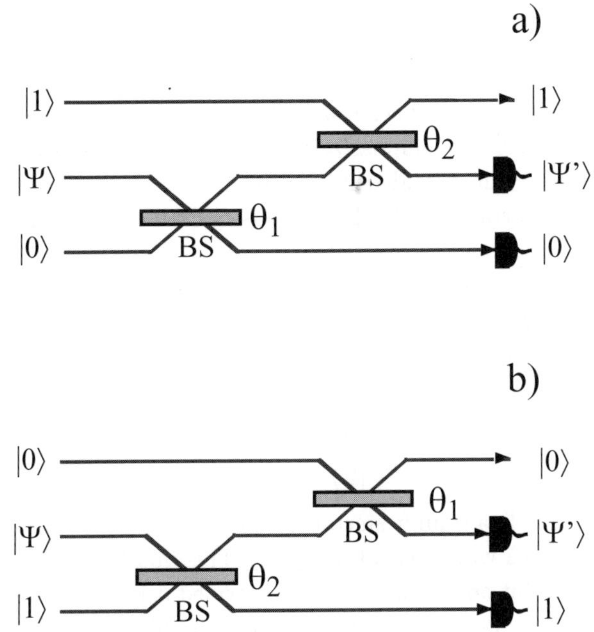

Fig. 3. Schematic diagram of the two variants of the KLM nonlinear sign shift gate. Both of these gates are composed of three input states, two beam-splitters and two photon number resolving detectors. Again the three input states are our unknown signal state $|\Psi\rangle = c_0|0\rangle + c_1|1\rangle + c_2|2\rangle$ plus the ancilla states $|1\rangle$ and $|0\rangle$ respectively. These three states interact with each other via the beam-splitters characterized by θ_1 and θ_2 (where $\cos^2\theta_1 = \frac{3-\sqrt{2}}{7}$ and $\cos^2\theta_2 = 5 - 3\sqrt{2}$). When a photon is detected in the first ancilla and the vacuum in the second ancilla, the signal state $|\Psi\rangle$ is transformed to $|\Psi'\rangle = c_0|0\rangle + c_1|1\rangle - c_2|2\rangle$. We observe that the sign of the $|2\rangle$ has flipped.

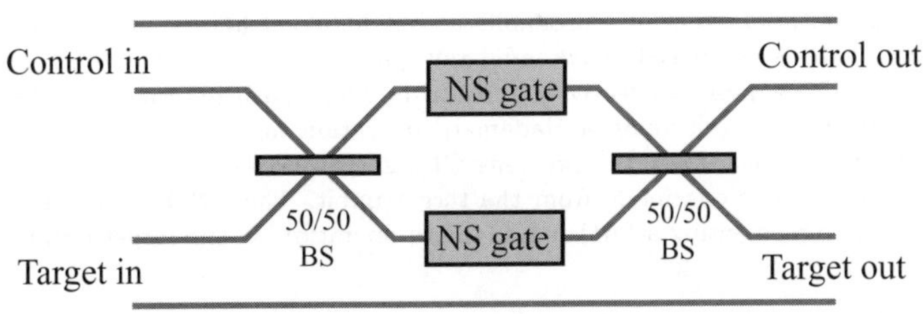

Fig. 4. Schematic diagram of a probabilistic but heralded linear optical controlled-phase (CPhase) gate. The CPhase gate is composed of two nonlinear sign shift gates (NS-gates) plus two 50/50 beam-splitters.

where we observe that the sign of the β_3 component has flipped due to the action of these gate. Finally applying the last 50/50 beam-splitter we have

$$\beta_0|00\rangle + \beta_1|01\rangle + \beta_2|10\rangle - \beta_3|11\rangle \tag{17}$$

which is what we would expect if a CPhase gate had been applied to our initial state. We must re-emphasize that this is not a deterministic gate. It is probabilistic but heralded in nature due to the two nonlinear sign shift gates used to construct it. The CPhase gate has a 1/16 probability of success.

In the above discussions on nonlinear sign shift and CPhase gates we have considered photon number states of light and not polarization states such as $|H\rangle$ and $|V\rangle$. It is well known that we can convert the latter to photon-number encoded states by a polarizing beam-splitter. Basically

$$|H\rangle \rightarrow |10\rangle \tag{18}$$

$$|V\rangle \rightarrow |01\rangle \tag{19}$$

We call this a dual rail qubit since the single mode polarization state has been converted to a which-path type qubit. This dual rail qubit is essential because single qubit operations can be performed using only beam-splitters and phase shifter. This is not the case for a qubit encoded in the basis states $|0\rangle$ and $|1\rangle$.

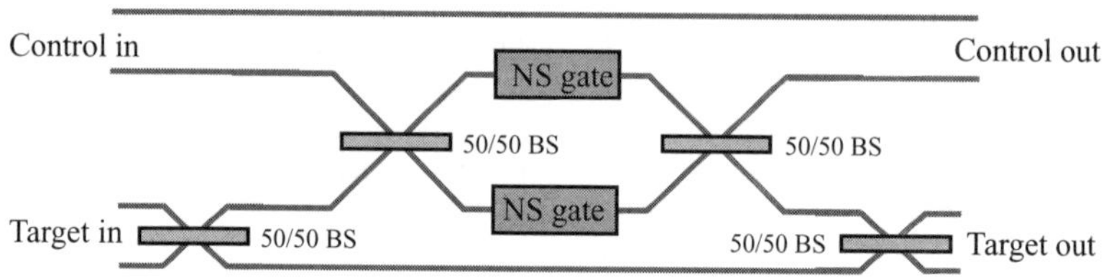

Fig. 5. Schematic diagram of a probabilistic but heralded linear optical CNOT gate. The CNOT gate is composed of two dual rail qubits (each qubit has the basis states $|10\rangle$ and $|01\rangle$). The qubit is effectively encoded in which-path information. The CNOT operation begins by performing a Hadamard operation on the target qubit utilizing a 50/50 beam-splitter. Then the previous CPhase gate is applied to one mode from the control qubit and one mode from the target qubit. The CNOT operation is then completed by implementing a further Hadamard operation on the target qubit.

This dual rail encoding then allows us a natural way to transform the CPhase gate to a CNOT gate (depicted schematically in Figure 5). The CNOT gate is basically the CPhase gate with a Hadamard operation being performed before and after it on the target qubit.

So far we have shown how it is possible to construct heralded but probabilistic two-qubit gates in linear optics. The CPhase and CNOT gates have a maximum probability of success of 1/16. A natural question is whether we can increase this probability of success. Recently Knill[20] has shown a slightly different CNOT circuit (not composed of nonlinear sign shift gates) which has an ideal probability of success of 2/27. Even with classical feed-forward it is not possible to significantly increase this. However, if we allow the use of entangled resources the probability of success for a CNOT gate for instance can be increased to one quarter[21]. The circuit is depicted in Figure 6.

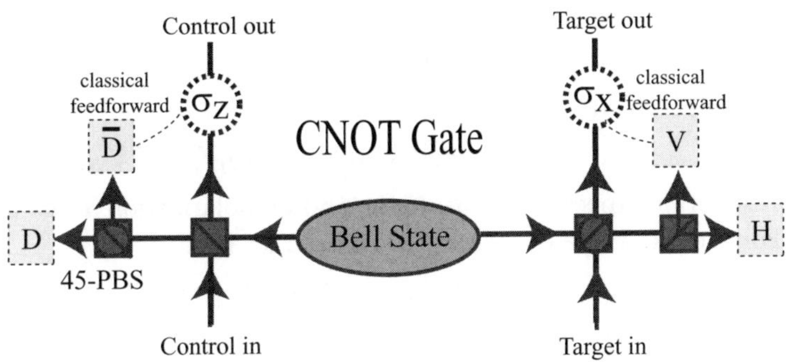

Fig. 6. Schematic diagram of Franson's four-photon CNOT gate. It is composed of two single-polarization encoded qubits (the control and target) plus one maximally entangled Bell state of the form $|HH\rangle + |VV\rangle$, plus polarizing beam-splitters (operating either in the $\{H, V\}$ or $\{D = H + V, \bar{D} = H - V\}$ bases) and two photon number resolving detectors. When the ancilla photons are detected at the LHS and RHS detectors, a CNOT operation is induced onto the control and target qubits.

Let us consider the operation of this gate in more detail. We will assume that the control and target are initially in the state

$$c_0|HH\rangle_{ct} + c_1|HV\rangle_{ct} + c_2|VH\rangle_{ct} + c_3|VV\rangle_{ct} \tag{20}$$

with the ancilla qubits prepared in the maximally entangled state $|HH\rangle_{ab}+|VV\rangle_{ab}$. This four-qubit state can be written as

$$\begin{aligned} &c_0|HHHH\rangle_{cabt} + c_0|HVVH\rangle_{cabt} + \\ &c_1|HHHV\rangle_{cabt} + c_1|HVVV\rangle_{cabt} + \\ &c_2|VHHH\rangle_{cabt} + c_2|VVVH\rangle_{cabt} + \\ &c_3|VHHV\rangle_{cabt} + c_3|VVVV\rangle_{cabt} \, . \end{aligned} \tag{21}$$

The action of the LHS {H,V} polarizing beam-splitter between the control and ancilla mode a bunches the photons if the polarizations are different and anti-bunches them if they are same. This results in the four-mode state

$$\begin{aligned}&c_0|H,H,H,H\rangle_{cabt} + c_0|HV,0,V,H\rangle_{cabt} +\\ &c_1|H,H,H,V\rangle_{cabt} + c_1|HV,0,V,V\rangle_{cabt} +\\ &c_2|0,VH,H,H\rangle_{cabt} + c_2|V,V,V,H\rangle_{cabt} +\\ &c_3|0,VH,H,V\rangle_{cabt} + c_3|V,V,V,V\rangle_{cabt}\end{aligned} \tag{22}$$

where the commas have been inserted to make it explicit which photons are in which mode. As a part of our protocol, we are going to condition the control and target qubits evolution on a single photon being detected in mode a. The protocol is aborted if zero or two photons are detected. Hence the above state conditioned by a single-photon detection in mode a is

$$\begin{aligned}c_0|H,H,H,H\rangle_{cabt} &+ c_1|H,H,H,V\rangle_{cabt}\\ &+ c_2|V,V,V,H\rangle_{cabt} + c_3|V,V,V,V\rangle_{cabt}\,.\end{aligned} \tag{23}$$

For the detection process we measure the photon polarization in the $\{D,\bar{D}\}$ basis. For a click at the D detector in mode a we have

$$c_0|H,H,H\rangle_{cbt} + c_1|H,H,V\rangle_{cbt} + c_2|V,V,H\rangle_{cbt} + c_3|V,V,V\rangle_{cbt} \tag{24}$$

while for a $\bar{D}$ click we have

$$c_0|H,H,H\rangle_{cbt} + c_1|H,H,V\rangle_{cbt} - c_2|V,V,H\rangle_{cbt} - c_3|V,V,V\rangle_{cbt}\,. \tag{25}$$

A sign flip (via classical feedforward) on the V polarization of the control qubits transforms the state (25) to (24). There is only a 50 percent chance that we will detect a single photon in ancilla a. The second part of the protocol is now to have the second ancilla qubit b and the target qubit interact on a polarizing beam-splitter operating in the $\{D,\bar{D}\}$ basis. Before we do this it is useful to rewrite the ancilla b and target qubit of (25). The state (25) then looks like

$$\begin{aligned}&[(c_0+c_1)\,|H\rangle_c + (c_2+c_3)\,|V\rangle_c]\,|DD\rangle_{bt}\\ &+[(c_0-c_1)\,|H\rangle_c - (c_2-c_3)\,|V\rangle_c]\,|\bar{D}\bar{D}\rangle_{bt}\\ &+[(c_0-c_1)\,|H\rangle_c + (c_2-c_3)\,|V\rangle_c]\,|D\bar{D}\rangle_{bt}\\ &+[(c_0+c_1)\,|H\rangle_c - (c_2+c_3)\,|V\rangle_c]\,|D\bar{D}\rangle_{bt}\,.\end{aligned} \tag{26}$$

In this form we can see that the $|DD\rangle_{bt}$ and $|\bar{D}\bar{D}\rangle_{bt}$ have the same polarization and hence will not bunch on the polarizing beam-splitter. The terms $|D\bar{D}\rangle_{bt}$ and $|D\bar{D}\rangle_{bt}$ are of opposite polarization and do bunch. If we require that one photon is detected in mode b then only the terms

$$\begin{aligned}&[(c_0+c_1)\,|H\rangle_c + (c_2+c_3)\,|V\rangle_c]\,|DD\rangle_{bt}\\ &+[(c_0-c_1)\,|H\rangle_c - (c_2-c_3)\,|V\rangle_c]\,|\bar{D}\bar{D}\rangle_{bt}\end{aligned} \tag{27}$$

can contribute. Now performing a measurement of the ancilla mode in the $\{H, V\}$ basis, our control and target qubits are transformed to

$$c_0|HH\rangle_{ct} + c_1|HV\rangle_{ct} + c_2|VV\rangle_{ct} + c_3|VH\rangle_{ct} \tag{28}$$

for an H result, and

$$c_0|HV\rangle_{ct} + c_1|HH\rangle_{ct} + c_2|VH\rangle_{ct} + c_3|VV\rangle_{ct} \tag{29}$$

for a V result. The second case can be transformed to the first by a bit flip of target qubit. The state (28) is what we would expect if a CNOT operation was performed on the initial state (20). This CNOT operation is probabilistic (but heralded) and works 25 percent of the time. Unfortunately these probabilistic gates cannot be used directly for scalable quantum computation because the probabilistic nature would lead to an exponential decrease in the probability of the entire computation being successful.

2.2. *Qubit Teleportation*

A very elegant solution to this problem has been proposed by Gottesman and Chuang[22] who have shown that quantum teleportation[23] can be used as a universal quantum primitive. In essence, quantum teleportation allows for the fault-tolerant implementation of "difficult" quantum gates that would otherwise corrupt the fragile information of a quantum state. The Gottesman and Chuang gate teleportation scheme[22] is depicted in Figure 7.

Let us consider this teleportation primitive in some detail. Suppose that we want to apply a CNOT operation between a control and target qubit. We assume that our control and target are initially in the state (20). We begin by creating two Bell states of the form $|HH\rangle + |VV\rangle$. We write their product in the form

$$|HHHH\rangle_{abef} + |HHVV\rangle_{abef} + |VVHH\rangle_{abef} + |VVVV\rangle_{abef} \,. \tag{30}$$

A CNOT operation is performed between the b and e qubits. When the gate is successful this operation creates the four qubit entangled state

$$|HHHH\rangle_{abef} + |HHVV\rangle_{abef} + |VVVH\rangle_{abef} + |VVHV\rangle_{abef} \,. \tag{31}$$

Now consider that a Bell measurement is performed between the control qubit and ancilla qubit a and between the target qubit and ancilla qubit f. From each Bell measurement we get one of the four results $|HH\rangle + |VV\rangle$, $|HV\rangle + |VH\rangle$, $|HH\rangle - |VV\rangle$ or $|HV\rangle - |VH\rangle$. The last three Bell states are related to the first by an X, Z or XZ operation. Hence we can specify the Bell-state measurement outcomes as I, X, Z or XZ. Consider that both

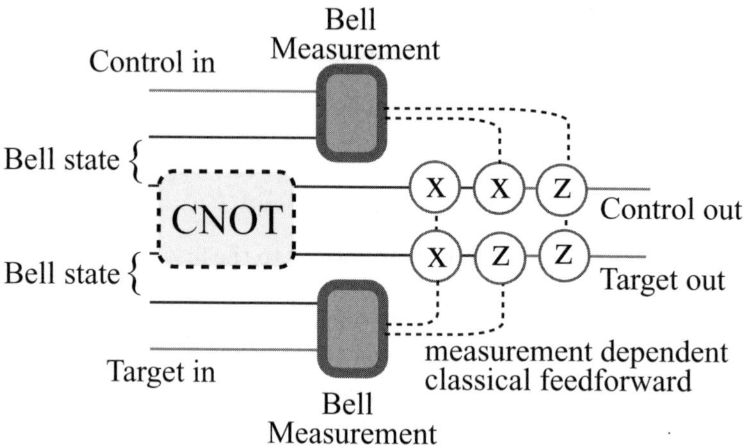

Fig. 7. Schematic diagram illustrating the teleportation of a two qubit gate. The scheme is composed of our initial control and target qubit plus two ancilla maximally entangled Bell states. The Bell states are entangled by the CNOT operation. This operation does not need to be deterministic, having a heralded result is sufficient. Only after the CNOT operation is successful, a Bell measurement is performed between the control and one of the qubits of the first Bell state and another between the target qubit and one of the qubits of the second Bell state. Classical feed forward operations perform bit/phase flips on the teleported qubits depending on the result of these measurements (as detailed below). This generates a CNOT operation between the control and target qubits.

Bell-state measurements give us the result I (that is, the $|HH\rangle+|VV\rangle$ state is detected). In this case the ancilla qubits b and e are transformed to

$$c_0|HH\rangle_{be} + c_1|HV\rangle_{be} + c_2|VH\rangle_{be} + c_3|VV\rangle_{be} \tag{32}$$

which is the state we would obtain had we applied the CNOT gate directly to the initial state (20). Similarly if the result of the Bell state measurements is X for the first measurement and I for the second we would obtain the state

$$c_0|VH\rangle_{be} + c_1|VV\rangle_{be} + c_2|HH\rangle_{be} + c_3|HV\rangle_{be} \tag{33}$$

which can be transformed to (32) by applying a bit flip (an X operation) to the first qubit b. Figure 7 indicates the classical feedforward operations needed to transform the states from all the Bell measurement results to (32). This really shows the power of gate teleportation. As long as we are able to perform Bell measurements efficiently, it is possible to transform probabilistic but heralded gates into deterministic ones. The key is that we keep applying the probabilistic gate to a known set of entangled states until it works. This creates a new entangled state. We then teleport the computational qubits using this new entangled state, and the computational qubits we get out have the appropriate gate operation applied to them. This is a very powerful technique and is in fact a universal quantum primitive.

Unfortunately, using only linear optics it is impossible to implement a Bell state measurement that distinguishes all four Bell states. With a single beam-splitter only two of the Bell states can be distinguished, and the scheme effectively works only 50 percent of the time. With two Bell-state measurements required for teleporting the CNOT gate the probability of success would be 1/4. To increase the success probability further we need to change the teleportation procedure[17]. Consider an initial entangled resource of the form

$$|T_n\rangle = \sum_{j=0}^{n} |1\rangle^j |0\rangle^{n-j} |0\rangle^j |1\rangle^{n-j}, \tag{34}$$

with $|a\rangle^j = |a\rangle_1 \otimes |a\rangle_2 \otimes \ldots \otimes |a\rangle_j$. This state $|T_n\rangle$ encodes n qubits with the k'th qubit located in the modes k and $n+k$. The teleportation protocol using the state $|T_n\rangle$ teleports mode 0 in a superposition of $|0\rangle$ and $|1\rangle$ to one of the last n modes of our entangled resource $|T_n\rangle$. This occurs by a Bell measurement implemented using an $n+1$ point Fourier transform implementable with passive linear optics and measurements of the number of photons in each of the modes $0 \ldots n$ on the mode to be teleported and the first n modes of $|T_n\rangle$. If the result of the measurement is $n = j$ then the teleported state is in the mode $n + j$. If the measurement result is 0 or $n + 1$, the Bell state measurement fails. This failure occurs with a probability $1/(n+1)$ and so the probability of success for the teleportation is now $n/(n+1)$. For $n = 1$ we get our well known $P = 1/2$. However as n increases significantly this probability approaches unity. Given that we now have an efficient teleportation protocol let us return to the teleportation of gates. In this instance we will consider the CPhase gate instead of the CNOT.

Consider now two copies of the state $|T_n\rangle$. These can be written in the form

$$|T_n\rangle|T_n\rangle = \sum_{j=0}^{n}\sum_{k=0}^{n} |1\rangle^j|0\rangle^{n-j}|0\rangle^j|1\rangle^{n-j}|1\rangle^k|0\rangle^{n-k}|0\rangle^k|1\rangle^{n-k} \tag{35}$$

If we wish to teleport a CPhase gate then we need to apply a CPhase operation to each pair of modes $(n + j, 3n + k)$ with j and k in the range $1 \ldots n$. When this operation is successful (remembering that the CPhase is heralded) we obtain the highly entangled state

$$|T'_n\rangle = \sum_{j,k=0}^{n} (-1)^{(n-j)(n-k)} |1\rangle^j |0\rangle^{n-j} |0\rangle^j |1\rangle^{n-j} |1\rangle^k |0\rangle^{n-k} |0\rangle^k |1\rangle^{n-k} \quad (36)$$

The teleportation protocols are applied between the control mode and modes $1 \ldots n$ and between the target mode and modes $3n+1 \ldots 4n$. Now given that we get measurements results (j,k) until sign flip corrections, the CPhase gate has been teleported on the modes $(n+j, 3n+k)$. The probability of success now scales as $n^2/(n+1)^2$ due to the two Bell measurements.

With this teleportation protocol we now have a mechanism for teleporting gates with near unit probability of success and hence we can achieve a universal set of gates (deterministic single qubits operation and the teleported CPhase or CNOT). The only disadvantage to this scheme is that the resource cost in generating the entangled state $|T'_n\rangle$ is high and potentially requires a very large number of single photons. This resource cost is polynomial and hence efficient. It does however raise the question of whether there is a more resource-efficient mechanism for single-photon linear optical quantum computation.

2.3. *One-way quantum computation*

An alternative approach to the usual gate based quantum computation schemes is the so-called cluster state or one-way quantum computation schemes. Using single photons, the cluster-state linear optical approach has a number of significant advantages including that it removes the need for the teleportation of non-deterministic gates[24,25]. Hence this is much more efficient in terms of physical resources. There have been several proposals to generate linear optical cluster states. The technique we will discuss here was first proposed by Browne and Rudolph[25].

The core elements in the Browne and Rudolph cluster state scheme are two "fusion" mechanisms. These allow for the construction of entangled photonic states. The core resources required are a source of maximally entangled two qubit photonic states. These can either be constructed with the linear optical techniques described in the previous subsection (we only need a heralded result) or via a two photon entangled source. Given the Bell states, we can proceed to build up the cluster states using only non-deterministic parity-check measurements, which involve combining the photons on a polarizing beam-splitter (PBS) followed immediately by measurement on the output modes.

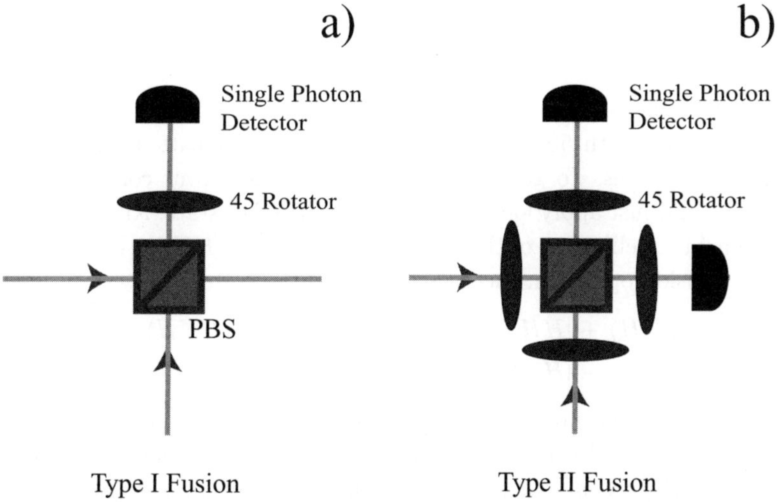

Fig. 8. Schematic diagram of two non-deterministic "qubit fusion" gates. Both gates are based on polarizing beam splitters (PBS). The Type I gate (subfigure (a)) consists of combining two spatial modes on a PBS, and measuring one of the output modes with a polarization discriminating photon counter after a 45° polarization rotation. The Type-II fusion gate (subfigure (b)) is constructed from the Type I gate by adding both 45° rotations to each input mode and measuring the output modes in the rotated basis.

Browne and Rudolph introduced two fusion gates depicted in Figure 8. Both of these gates are based on Franson parity-check operation. Let us consider the action of the Type I "qubit fusion" gate. This operation by combining our two modes of interest on a polarizing beam splitter (PBS) and then rotating one of the output modes by 45° before measuring it with a polarization discriminating photon counter. Consider two Bell states of the form $|HH\rangle + |VV\rangle$. We can write these in the form

$$|HH\rangle_{12}|HH\rangle_{34} + |HH\rangle_{12}|VV\rangle_{34} + |VV\rangle_{12}|HH\rangle_{34} + |VV\rangle_{12}|VV\rangle_{34} \quad (37)$$

Now taking the modes 2 and 3 and combining them on the PBS we get

$$|HH\rangle_{12}|HH\rangle_{34} + |VV\rangle_{12}|VV\rangle_{34} \quad (38)$$

where we have kept only the terms in which one photon remains in each mode (this occurs 50 percent of the time). Now measuring in the $\{D, \bar{D}\}$ basis of the second mode we get upon the detection of one and only one photon

$$|HHH\rangle_{134} + |VVV\rangle_{134} \quad (39)$$

for the D result and

$$|HHH\rangle_{134} - |VVV\rangle_{134} \tag{40}$$

for $\bar{D}$. It is simple to convert the case (40) to (39), a maximally entangled GHZ state. Thus with 50 percent probability we can create a GHZ state from two Bell states. To generate three-qubit clusters (our core building block), we could just apply local operations to the GHZ state. Alternatively we can consider Bell states of the form $|HH\rangle + |VH\rangle + |HV\rangle - |VV\rangle$. After the Type I fusion operation succeeds we obtain the state

$$\begin{aligned} &|HHH\rangle + |HHV\rangle + |VHH\rangle + |VHV\rangle + |HVH\rangle \\ &\qquad - |HVV\rangle - |VVH\rangle + |VVV\rangle \end{aligned} \tag{41}$$

If this Type-I fusion gate is applied to the end-qubits of linear clusters (Figure 9(a)) of lengths n and m, a successful operation generates a linear cluster of length $(n+m-1)$. However for the Type-I fusion gate to operate successfully we need photon number resolving detectors. However, once we have generated the three-qubit clusters, we can join them via the Type-II fusion gate. Each time, with probability $1/2$, the cluster grows in length by 2 qubits, or, equally likely, loses a qubit. A failed attempt creates a Bell pair from the 3-qubit cluster, which can be re-used for the generation of further 3-qubit clusters. Hence, on average, the cluster grows by $1/2$ a qubit in length for each attempt, and the resources needed scale as $(2 \times 4 - 1) = 7$ Bell pairs per qubit in the linear cluster. With more efficient strategies this resource cost can be decreased slightly further.

It is however well known that one-dimensional cluster states are not sufficient for universal quantum computation. It is necessary as a minimum to create two-dimensional cluster states. This can be achieved using the Type-II fusion gate between the mid points of the linear chains (see Figure 9(b)). Thus it is possible to create arbitrary two dimensional cluster states and as arbitrary single-qubit measurements are easy to perform on photonic qubits we have an efficient mechanism to perform universal quantum computation.

2.4. *Nonlinear optics*

Single photon nonlinear optics is very similar to the linear optical schemes described above, but uses other resources (rather than single photons, beam-splitters and single photon sources) to generate the nonlinear effects on the optical modes. In the linear optics schemes the nonlinear sign shift gate and CNOT (or equivalently CZ) gates were performed in a heralded

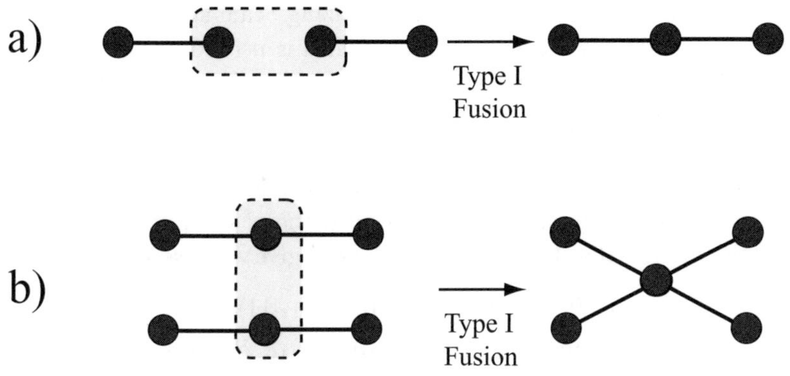

Fig. 9. Schematic diagram showing the fusing of two cluster states. In the first situation a) we fuse the end qubits of two linear clusters of length n and m. This creates a cluster of length $(n + m - 1)$. In the second case b) the mid qubits of two linear clusters are fused to create a two-dimensional cluster with a cross-like layout. In this way non-trivial cluster layouts can be created.

but probabilistic fashion. This means that the teleportation trick was required to make the gates deterministic and hence significantly increased the required resources. It has long been known that a strong cross-Kerr nonlinearity is sufficient to create the CNOT gate (CZ, in fact, which may be converted to CNOT). However, natural materials do not currently process the massive, reversible nonlinearities required. Nevertheless, it is possible to engineer atomic systems to create large effective optical nonlinearities and it is these we will discuss briefly in this section. There are two regimes which are particularly important in using atomic systems to generate optical nonlinearities:

- Passive optical nonlinearities[26,27],
- Near deterministic heralded optical nonlinearities[28].

Let us begin by examining the passive optical nonlinearities created by atomic systems. By this we mean nonlinearities generated without the need for active measurement. Electromagnetically induced transparency (EIT) which has recently been demonstrated in a number of situations is an ideal candidate for realizing controlled phase gates using the giant Kerr nonlinearities available[26,27]. If we consider the four level (N) configuration depicted in Figure 10 below, then we can send the control mode into mode a and the target mode into mode c. The qubits could be encoded into which-path or dual rail photons. In the latter case, the polarization encoding can be turned into dual rail encoding using polarizing beam-splitters, as shown in Figure 10. Mode b is a classical pump, which is needed to link modes a and c. Such an EIT system generates an effective nonlinearity of the form

$$U = e^{-i\chi t \hat{n}_a \hat{n}_c} \tag{42}$$

where χ is the strength of the nonlinearity and t is the interaction time. Only when there is a photon present in both the a and c modes passing through the atoms (the $|11\rangle$ amplitude) does U give a phase shift, that is

$$\begin{aligned} U \quad & (c_{00}|00\rangle + c_{01}|01\rangle + c_{10}|10\rangle + c_{11}|11\rangle) \\ &= \left(c_{00}|00\rangle + c_{01}|01\rangle + c_{10}|10\rangle + c_{11}e^{-i\chi t}|11\rangle\right) \ . \end{aligned} \tag{43}$$

By choosing χt appropriately we can tune such that $e^{-i\chi t} = -1$ and hence generate a controlled phase shift of π. This controlled phase shift together with local qubit operations is sufficient to generate a CNOT gate.

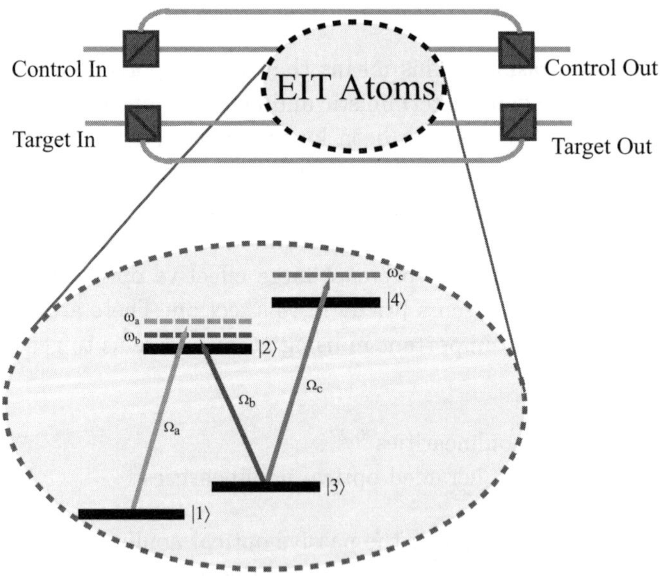

Fig. 10. Atomic systems (with a N configuration of levels) can be used to produce a CZ gate. The atoms provide an effective cross-Kerr nonlinearity, so passage of photons through the two rails that interact with the atoms (the $|11\rangle$ amplitude) will realize a phase shift that can be adjusted to π, to give a sign change relative to the other amplitudes. If the photons are initially encoded in polarization, this can be converted into which-path or dual rail encoding through polarizing beam-splitters. These send one polarization amplitude, representing $|0\rangle$, around a path avoiding the atoms, and the other, representing $|1\rangle$, through the region containing the atoms.

Next let us consider the generation of heralded optical nonlinearities in a near deterministic fashion. The real key here is to use one of the key advantages from the all linear optical schemes, that is the conditioning measurement. However in these atomic cases we want the conditioning to be near-deterministic (with a success probability near to unity). An elegant scheme has been proposed by Gilchrist et al.[28], which allows a nonlinear sign shift (NS) gate to be implemented with a probability exceeding 99 percent. Two of the NS gates can be used to build a CNOT gate. For this situation the conditioning measurement is done via a highly efficient ion-trap shelving measurement and it is the efficiency of this measurement that makes the overall gate near-deterministic. As the gate is near-deterministic it is possible that the teleportation trick will not be needed and hence the overall resources required for the information processing will be less. This is the advantage of using heralded measurements to induce optical nonlinearities.

So far we have considered a number of mechanisms to perform universal single-photon quantum computation. However as we have mentioned we can use other states of light, not restricted to qubits. We could perform qudit computation, but more importantly continuous variable computation is also available and this is what we will consider next.

3. Continuous variables

As we have discussed, a qubit lives in a two-dimensional Hilbert space. However, some quantum systems live in larger spaces—even infinite-dimensional Hilbert spaces, such as for a particle moving in one dimension. Such a particle can be characterized by its position X and momentum P. In the context of quantum information, such a system is termed a qunat - the variables X and P have a continuous spectrum. Clearly, then, continuous variables are not restricted to the realm of optics. However, to date, the only continuous variables that have been successfully employed in quantum information are optical. Here X and P represent the two quadratures of an electromagnetic field mode (or the electric and magnetic fields).

Given such a significant change in the way we encode our information, we need to carefully examine what operations are possible in these continuous variables systems and whether they could be universal. At first it may seem that quantum computation over continuous variables is an ill-defined concept. What does it mean to perform computation with continuous variables? First, there are an infinite number of parameters (each for a single particle or mode). Do we need to manipulate them all?

These are not easy questions to answer, but significant insight can be achieved by reconsidering universal computation with qubits. A qubit quantum computer could be defined as a device that applies local operations via quantum logic gates that affect only a few variables of this device at a time. By repeated application of these logic gates (for instance Hadamard, $\pi/8$ and the CNOT) we can effect any unitary transformation over a finite number of those variables to any desired degree of precision. So what does this mean in the continuous variable case? An arbitrary unitary transformation over even a single continuous variable requires an infinite number of parameters to define and typically cannot be approximated by a finite number of continuous quantum operations. However, it is possible to define the notion of universal quantum computation over continuous variables for various subclasses of transformations. One well considered case is the set of operations that correspond to Hamiltonians which are polynomial functions of the operators X and P of the continuous variables[29]. This set of continuous variable operations can be termed universal for the particular set of transformations if one can, through a finite number of applications of the operations, approach arbitrarily closely to any transformation in that set. However, the set cannot be too simple, or it could be efficiently classically simulated[11].

Let us start with the construction of a linear Hamiltonian of X and P of the form $aX + bP + c$. Basically this Hamiltonian can be constructed by applying the X operator for a short time adt, then P for a time bdt, followed by the operators P, X, $-P$ and X for a time $\sqrt{c}dt$. Using the relation

$$e^{iAdt}e^{iBdt}e^{-iAdt}e^{-iBdt} = e^{i[A,B]dt+O(dt^3)} \tag{44}$$

we can see that the net effect of these operations is our desired transformation $\exp\left[i(aX + bP + c)dt\right]$. Now by making dt sufficiently small, one can approach arbitrarily close to effecting a Hamiltonian of the desired form over small times. Then, by repeating the small-time construction t/dt times, one can approach arbitrarily close to effecting the desired Hamiltonian over time t. Thus we have a mechanism to construct any linear Hamiltonian in X and P.

Next let us consider the construction of arbitrary quadratic Hamiltonians. Suppose now that one has the two quadratic Hamiltonians $H = (X^2 + P^2)$ and $S = (XP + PX)$. The Hamiltonian H is proportional to the energy operator and S is the squeezing operator. It is straightforward to show that any quadratic Hamiltonian in X and P can be constructed from repeated applications of H, S, X and P using similar techniques to the linear Hamiltonian case. The action of the operators H, S, X and P in sequence only results in Hamiltonian terms that are at most quadratic. It is impossible from these operators to generate higher order terms.

To construct higher order Hamiltonians, nonlinear operations such as the Kerr Hamiltonian $H^2 = (X^2 + P^2)^2$ are required. The Kerr Hamiltonian is a natural choice, but any nonlinear operation is sufficient. These higher order Hamiltonians have the key feature that commuting them with any Hamiltonian of the form H, S, P, X results in a Hamiltonian higher in order in X or P. For instance, the commutation between H^2 and H, S, X and P allows the construction of any cubic Hamiltonian. Then the commutation between H^2, H, S, X, P and these cubic Hamiltonians allows the construction of any fourth order Hamiltonian, and so on. By an iterative procedure one can construct single mode Hamiltonians that are arbitrary polynomials of X and P.

However for universal quantum computation we need a multi-mode system. To construct multi-mode Hamiltonians we just require beam-splitters plus arbitrary single mode polynomials of X and P. Since the output modes of the beam-splitters are superpositions of the input modes, this allows the construction of multi-mode Hamiltonians in X and P for the respective modes.

The key to this form of universal computation is the nonlinear operation and how it can be realized. Potentially this can be done using an EIT medium, or via heralded probabilistic operations and gate teleportation. In the heralded case we could use the NS gate discussed in the linear optical schemes, but with a more general signal mode and continuous variable teleportation rather than qubit teleportation. Without such nonlinear operations, our entire computation could be efficiently simulated classically.

3.1. *Continuous variables quantum gate teleportation*

We have previously found that qubit gate teleportation is an essential resource for single photon optical quantum computation. Continuous variable teleportation can also be used in this role, but has the significant advantage that it can be used to teleport both qubit, qudit and continuous-variable (CV) states and gates. Before we consider continuous variable gate teleportation let us review CV teleportation. Consider a three mode optical system with the modes a and b being the maximally entangled Einstein-Podolsky-Rosen (EPR) state $|\mathrm{EPR}\rangle_{ab} = \int |q\rangle_a |q\rangle_b \, dq$ and mode c being an arbitrary pure state $|\psi\rangle$. In the EPR state $|q\rangle$ are the position eigenstates. The teleportation protocol now works as follows: Modes a and c are subjected to the transformation joint projective measurements

$$\Pi_{q,p} = R_c(q,p)|\mathrm{EPR}\rangle_{ac}\langle\mathrm{EPR}|R_c^\dagger(q,p), \tag{45}$$

where $R_c(x,p)$ is the Pauli operator

$$R(q,p) = \exp\left[-i(q\hat{p} - p\hat{q})\right] \tag{46}$$

on mode c. Specific examples of this operator are $R(q,0) = X(q)$ and $R(0,p) = Z(p)$. Now this measurement yields two classical numbers, q_0 and p_0. These are then used to condition the Pauli operation $R_b(q_0,p_0)$, after which mode b is left in the state $|\psi\rangle$. We can thus say that the state has been teleported from mode c to mode b. This operation is schematically presented in Figure 11.

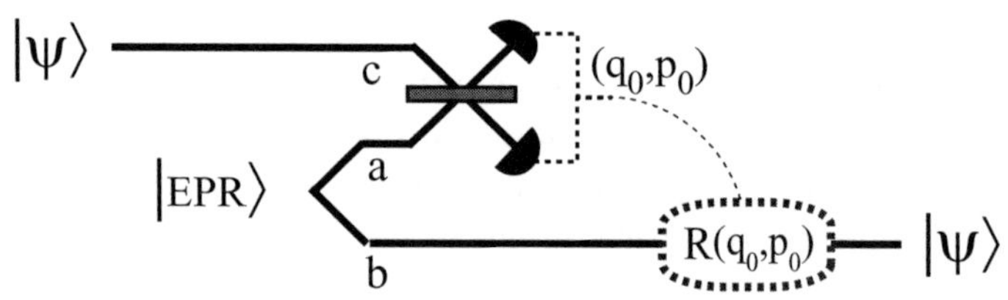

Fig. 11. Schematic diagram of continuous variable quantum teleportation of state $|\psi\rangle$ from mode c to mode b.

We can now proceed to look at the teleportation of gate (depicted schematically in Figure 12). The process of first teleporting the state $|\psi\rangle$ and then implementing the gate U is equivalent to acting on one mode b of the EPR state with U, followed by a modified quantum teleportation. There are a number of U gates that we can consider but as a start let U be in the single mode Clifford group. In this case we know that

$$UR(q_0,p_0) = R'(q_0,p_0)U, \tag{47}$$

where $R'(q_0,p_0) = UR(q_0,p_0)U^{-1}$ is also an element of the Pauli group. Thus, the desired Clifford gate U can be quantum teleported onto the state $|\psi\rangle$ simply by implementing U on one mode of the EPR pair and by appropriately altering the conditional displacement of the quantum teleportation (see Figure 12). By extension, using n single-mode quantum teleportation circuits, it is possible to teleport any gate in the n-mode Clifford group.

This scheme can be used to teleport more general quantum gates than those in the Clifford group. Once we move to the more general gates we are

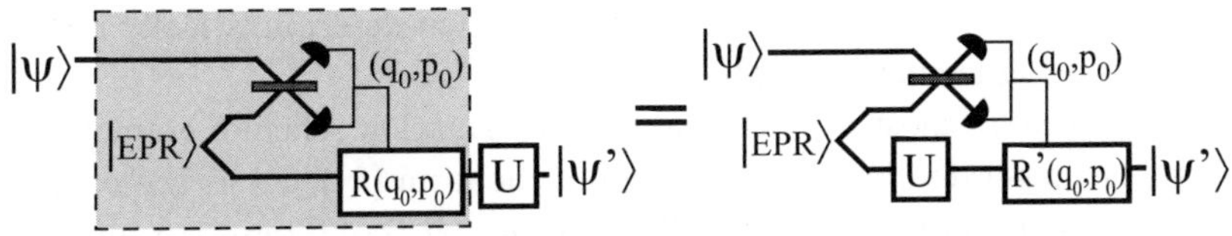

Fig. 12. Schematic diagram of continuous variable quantum teleportation of a gate U. The process of first teleporting the state $|\psi\rangle$ and then implementing the gate U, resulting in the state $|\psi'\rangle = U|\psi\rangle$ is equivalent to acting on one mode of the EPR state with U, followed by a modified quantum teleportation.

using nonlinear transformations and it is these nonlinear operations that enable universal quantum computation. For example, such a transformation could be the *cubic phase gate*

$$V(\gamma) = \exp(\mathrm{i}\gamma\hat{q}^3) \; . \tag{48}$$

Let us consider the teleportation of this gate. To proceed we need to know how to commute this gate back through the Pauli operators $R(q_0, p_0)$. We observe that

$$V(\gamma)R(q_0, p_0) = R_2'(q_0, p_0, \gamma)V(\gamma) \; , \tag{49}$$

where

$$\begin{aligned} R_2'(q_0, p_0, \gamma) &= V(\gamma)R(q_0, p_0)V(\gamma)^{-1} \\ &= \exp\left[-\mathrm{i}\big(q_0\hat{p} - p_0\hat{q} - 3\gamma q_0\hat{q}^2\big)\right] \; . \end{aligned} \tag{50}$$

The term $R_2'(q_0, p_0, \gamma)$ is quadratic in q, p and so can be implemented with our Clifford group operations (displacements and squeezing). Thus, to teleport $V(\gamma)$, this nonlinear gate is performed on the second mode of the EPR state. We follow this by a modified teleportation scheme with a conditional operation $R_2'(q_0, p_0, \gamma)$ on the b mode. The result is that a state $|\psi\rangle$ is teleported into the transformed state $V(\gamma)|\psi\rangle$. Now the cubic phase gate $V(\gamma)$, being a higher-order nonlinear gate on a single mode, can be combined with Clifford group gates of n modes to form a universal set of gates for QC on n modes. A scheme is known to implement this cubic phase gate, and thus with continuous variable quantum teleportation (CVQT) it is possible to teleport a universal and realizable set of gates. If the Clifford group transformations and CVQT can be implemented fault-tolerantly, then it is possible to use this scheme to implement a fault-tolerant cubic phase gate using a gate that is not fault-tolerant. This is the key result and shows that any nonlinear transformations can be moved "off-line". Although these transformations must still be performed in the quantum teleportation circuit, they can be made to act on EPR ancilla states non-deterministically rather than on the fragile encoded states.

Lastly because most optical quantum information schemes employ the Kerr effect (generated by a Hamiltonian of the form $(\hat{a}^\dagger)^2\hat{a}^2$) as the nonlinear transformation outside of the Clifford group, it is of interest to consider how such a transformation can be implemented using the above universal set of gates. Using the relation $e^{\mathrm{i}At}e^{\mathrm{i}Bt}e^{-\mathrm{i}At}e^{-\mathrm{i}Bt} = e^{\mathrm{i}[A,B]t^2} + \mathcal{O}(t^3)$, a combination of cubic phase gates and Clifford gates can be used to simulate the Kerr nonlinearity to any degree of accuracy.

4. Hybrid schemes: combining single photon logic and continuous variables

Now that we have considered quantum computation using single photons and continuous variables, let us examine hybrid schemes where we use elements from both. Our primary motivation here is to encode the quantum information using the single photons and to use the continuous variables as the communication channel between the qubits. The key elements in this hybrid scheme are weak cross-Kerr nonlinearities, too weak to realise a CPhase gate directly.

4.1. *A QND detector*

Before we begin our discussion of the construction of efficient quantum gates using weak nonlinearities, let us first consider a photon number quantum non-demolition (QND) measurement using a cross-Kerr nonlinearity[30]. The cross-Kerr nonlinearity has a Hamiltonian of the form

$$H_{QND} = \hbar\chi\hat{n}_a\hat{n}_c \tag{51}$$

with the signal (probe) mode having the photon number operators $\hat{n}_a = a^\dagger a$ $(\hat{n}_c = c^\dagger c)$ where $a^\dagger, a$ $(c^\dagger, c)$ are the respective creation and annihilation operators for the modes and χ is the strength of the nonlinearity. This Hamiltonian is photon-number preserving, that is, it does not change the photon number in either the signal or probe modes. This Hamiltonian enables a unitary transformation of the form

$$U_{QND} = \exp\left[-i\theta\hat{n}_a\hat{n}_c\right] \tag{52}$$

to be implemented where $\theta = \chi t$ with t being the interaction time between the signal and probe beams with the Kerr material.

Let us now consider its action on the signal and probe modes. If the signal field initially contains n_a photons and the probe field is in an initial coherent state with amplitude α_c, the cross-Kerr nonlinearity causes the combined system to evolve as

$$|\Psi(t)\rangle_{out} = U_{QND}|n_a\rangle|\alpha_c\rangle = |n_a\rangle|\alpha_c e^{in_a\theta}\rangle. \tag{53}$$

The Fock state $|n_a\rangle$ is unaffected by the interaction with the cross-Kerr nonlinearity but the coherent state $|\alpha_c\rangle$ picks up a phase that is directly proportional to the number of photons n_a in the signal state $|n_a\rangle$. If we measure this phase shift we can then indirectly infer the number of photons in the signal mode a. This can be achieved simply with a homodyne measurement (depicted schematically in Figure 13).

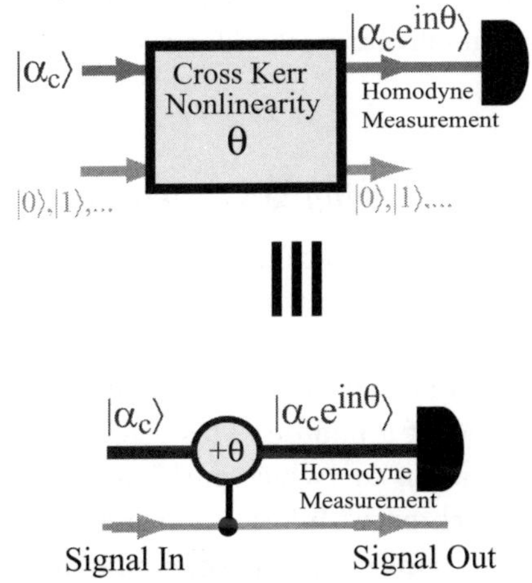

Fig. 13. Schematic diagram of a photon resolving detector based on a cross-Kerr nonlinearity and a homodyne measurement. The two inputs are a Fock state $|n_a\rangle$ (with $n_a = 0, 1, ..$) in the signal mode a and a coherent state with real amplitude α_c in the probe mode c. The presence of photons in mode a causes a phase shift on the coherent state $|\alpha_c\rangle$ directly proportional to n_a, which can be determined with a momentum quadrature measurement on mode c.

The homodyne apparatus allows measurement of the quadrature operator $x(\phi) = ce^{i\phi} + c^\dagger e^{-i\phi}$ with ϕ being the phase of the local oscillator. This has the expectation value

$$\langle x(\phi)\rangle = 2\mathrm{Re}\,[\alpha_c]\cos\delta \tag{54}$$

where $\delta = \phi + n_a\theta$. For a real and positive initial α_c, a highly efficient homodyne measurement of the position $X = c + c^\dagger$ or momentum $Y = \frac{c-c^\dagger}{i}$

quadratures yield the expectation values

$$\begin{aligned}\langle X\rangle &= 2\alpha_c \cos(n_a\theta)\\ \langle Y\rangle &= 2\alpha_c \sin(n_a\theta)\end{aligned} \tag{55}$$

with a unit variance. If the inputs in mode a are the Fock state $|0\rangle$ or $|1\rangle$, the respective outputs of the probe mode c are either the coherent states $|\alpha_c\rangle$ or $|\alpha_c e^{i\theta}\rangle$. The probability of misidentifying these states is given by

$$P_{\text{error}} = \frac{1}{2}\text{Erfc}\left[\frac{\text{SNR}_Y}{2\sqrt{2}}\right]. \tag{56}$$

where $\text{SNR}_Y = 2\alpha_c \sin(n_a\theta)$. A signal to noise ratio of $\text{SNR}_Y = 2\pi$ would thus give $P_{\text{error}} \sim 10^{-4}$ and hence we require $\alpha_c \sin\theta \approx \pi$. This can be achieved with a small nonlinearity θ as long as the probe beam is intense enough[30].

4.2. *A polarization-preserving QND detector*

In many optical quantum computation tasks our information is not encoded in photon number but in polarization instead. When our information encoding is polarization-based there are two separate detection tasks that we need to perform. The first and simplest is just to determine, for instance, the polarization ($|H\rangle$ or $|V\rangle$) of a photon. This can be achieved by converting the polarization information to "which-path" information on a polarizing beam-splitter. The "which-path" information is photon-number encoded in each path and hence a separate QND photon number measurement of each path will determine which polarization basis state the photon was originally in, or project it into this basis. This can be thought of as two applications of the QND detector.

The second task (one that is critically important for error correction codes in optics) is to determine whether our single-photon polarization-encoded qubit is present or not. That is, for the optical field under consideration we want to determine whether it contains a photon or not. If it does contain a photon, we do not want to destroy the information in its polarization state. This can be achieved by first converting the polarization qubit to a "which-path" qubit. Each path then interacts with a weak cross-Kerr nonlinearity θ using the same shared probe beam (Figure 14). If a photon is present in either path of this signal beam it induces a phase shift θ on the probe beam; however, with this configuration it is not possible to determine which path induced the phase shift.

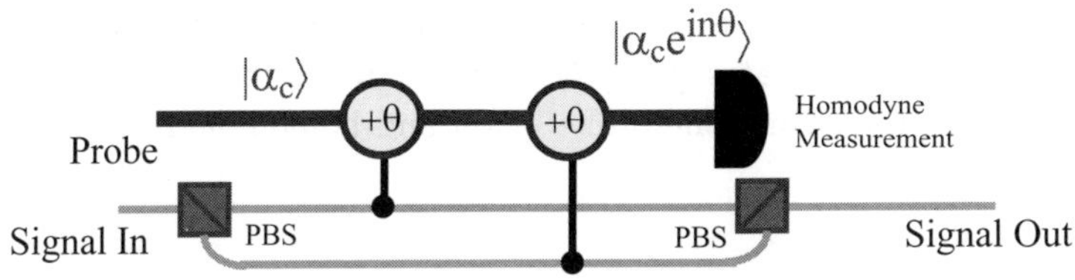

Fig. 14. Schematic diagram of a polarization-preserving photon number quantum nondemolition detector based on a pair of identical cross-Kerr optical nonlinearities. The signal mode is a Fock state with an unknown polarization, which is converted into which-path encoding by a polarizing beam splitter (square box). The phase shift suffered by the probe mode is proportional to n_a, independent of the polarization of the signal mode photon.

To illustrate this, consider a signal mode initially prepared in the state $c_0|0\rangle + c_H|H\rangle + c_V|V\rangle$. The action of the polarizing beam-splitter converts this polarization-encoded state to the which-path-encoded state $c_0|00\rangle + c_H|10\rangle + c_V|01\rangle$. Now the interaction between the path mode, probe field and the first nonlinearity creates the three mode state

$$c_0|00\rangle|\alpha\rangle_p + c_H|10\rangle|\alpha e^{i\theta}\rangle_p + c_V|01\rangle|\alpha\rangle_p \,. \tag{57}$$

The second weak nonlinearity causes a phase shift on probe beam component of the last term. This results in the three-mode system evolving to

$$c_0|00\rangle|\alpha\rangle_p + [c_H|10\rangle + c_V|01\rangle]\,|\alpha e^{i\theta}\rangle_p \,. \tag{58}$$

The last polarizing beam-splitter changes the which-path encoding back to the polarization encoding, resulting in

$$c_0|0\rangle|\alpha\rangle_p + [c_H|H\rangle + c_V|V\rangle]\,|\alpha e^{i\theta}\rangle_p \,. \tag{59}$$

It is now very clear that if a phase shift is detected, we know that a single photon is present in the signal mode without destroying its polarization encoding. The real key in this setup is the use of separate but identical cross-Kerr nonlinearities which allows different paths to interact with the same probe beam, each potentially inducing a phase shift on it. If a phase shift is detected we can infer the presence of a photon, but the coherence between the paths is perserved.

Finally it is enlightening to ask what happens if phase shifts from the different paths are not identical. If we consider the above example with the upper path inducing a phase shift θ and the lower path inducing $-\theta$ phase shift, then the resulting three-mode state is

$$c_0|0\rangle|\alpha\rangle_p + c_H|H\rangle|\alpha e^{i\theta}\rangle_p + c_V|V\rangle|\alpha e^{-i\theta}\rangle_p. \tag{60}$$

This is interesting because the homodyne measurement now allows us to distinguish all three states $|0\rangle$, $|H\rangle$ and $|V\rangle$ without having to use two full QND detectors. Only one probe beam and homodyne measurement is required.

4.3. *A two-qubit parity gate*

The concept of using multiple cross-Kerr nonlinearities is very interesting, as we have seen in the previous section. However, there is no reason to require the probe beam to interact only with one photonic qubit. This detection concept could be applied to multiple qubits. If we want to perform a more "generalized" type of measurement between different photonic qubits, we could delay the homodyne measurement, having the probe beam interact with several cross-Kerr nonlinearities where the signal mode is different in each case, as in Figure 15. The different signal modes could be from separate photonic qubits. The probe beam measurement then occurs after all these interactions, in a collective way which could, for instance, allow a non-destructive detection that distinguishes superpositions and mixtures of the states $|HH\rangle$ and $|VV\rangle$ from $|HV\rangle$ and $|VH\rangle$. The key here is that we could have no net phase shift on the $|HH\rangle$ and $|VV\rangle$ terms whilst having a phase shift on the $|HV\rangle$ and $|VH\rangle$ terms. We will call this generalization a *two-qubit polarization parity QND gate.*

Let us consider two polarization-encoded qubits given by

$$|\Psi_{12}\rangle = \beta_0|HH\rangle + \beta_1|HV\rangle + \beta_2|VH\rangle + \beta_3|VV\rangle\,. \tag{61}$$

We have written this as the most general two qubit pure state. Now consider the action of the first weak nonlinearity on the $|H\rangle$ component of the first qubit. The state of the system (both qubits and probe beam) is

$$\beta_0|HH\rangle|\alpha_c e^{i\theta}\rangle_p + \beta_1|HV\rangle|\alpha_c e^{i\theta}\rangle_p + \beta_2|VH\rangle|\alpha_c\rangle_p + \beta_3|VV\rangle|\alpha_c\rangle_p\,. \tag{62}$$

Next, application of the second cross-Kerr nonlinearity on the $|H\rangle$ component of the second qubit leads to

$$|\psi\rangle_T = [\beta_0|HH\rangle + \beta_3|VV\rangle]\,|\alpha_c\rangle_p + \beta_1|HV\rangle|\alpha_c e^{i\theta}\rangle_p + \beta_2|VH\rangle|\alpha_c e^{-i\theta}\rangle_p\,.$$

This state is illustrated in the phase-space plot in Figure 16. It is now obvious that the $|HH\rangle$ and $|VV\rangle$ terms pick up no phase shift and remain coherent with respect to each other while the $|HV\rangle$ and $|VH\rangle$ pick up opposite sign phase shifts of θ.

We thus need to perform a measurement that does not allow the sign of the phase shift to be determined. An X homodyne measurement achieves

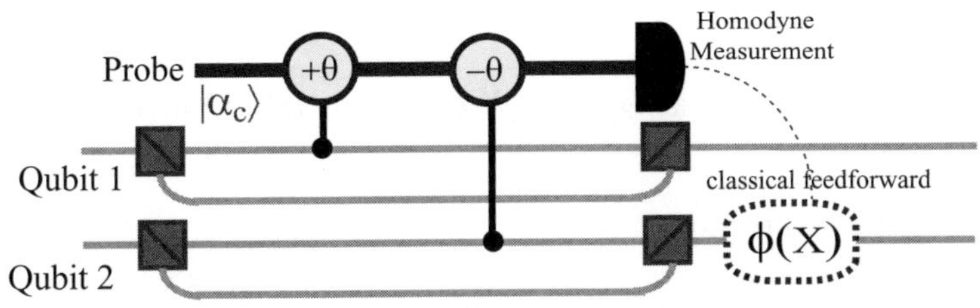

Fig. 15. Schematic diagram of a two-qubit polarization QND detector that distinguishes superpositions and mixtures of the states $|HH\rangle$ and $|VV\rangle$ from $|HV\rangle$ and $|VH\rangle$, using several cross-Kerr nonlinearities and a coherent laser probe beam $|\alpha_c\rangle$. The scheme works by first splitting each polarization qubit into a which-path qubit on a polarizing beam-splitter. The action of the first cross-Kerr nonlinearity puts a phase θ on the probe beam only if a photon is present in the upper path of the first qubit. The second cross-Kerr nonlinearity puts a phase $-\theta$ on the probe beam only if a photon is present in the upper path of the second qubit. After the nonlinear interactions the which-path encoded qubits are converted back to polarization encoded qubits. The probe beam only picks up a phase if the states $|HV\rangle$ and/or $|VH\rangle$ are present and hence the appropriate homodyne measurement allows the states $|HH\rangle$ and $|VV\rangle$ to be distinguished from $|HV\rangle$ and $|VH\rangle$. The two-qubit polarization QND gate thus acts like a parity checking device. If we suppose that the input state of the two polarization encoded qubits is $|HH\rangle+|HV\rangle+|VH\rangle+|VV\rangle$, then after the parity gate (using an X homodyne measurement) we have conditioned either to the state $|HH\rangle+|VV\rangle$, or to $e^{i\phi(X)}|HV\rangle+e^{-i\phi(X)}|VH\rangle$ where $\phi(X)$ is a phase shift that depends on the exact result of the homodyne measurement[31]. A simple phase shift achieved via classical feedforward then allows this second state to be transformed to the first if we wish.

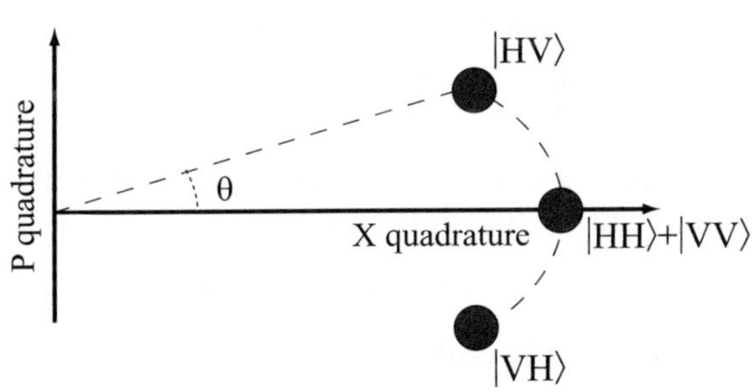

Fig. 16. Schematic phase-space illustration of the three mode quantum state $|\psi\rangle_T = [\beta_0|HH\rangle+\beta_3|VV\rangle]\,|\alpha_c\rangle_p+\beta_1|HV\rangle|\alpha_c e^{i\theta}\rangle_p+\beta_2|VH\rangle|\alpha_c e^{-i\theta}\rangle_p$. For the $|HH\rangle$ and $|VV\rangle$ components of the state, the probe beam does not suffer from a phase shift, while for the $|HV\rangle$ and $|VH\rangle$ components, the probe beam acquires phase shifts of θ and $-\theta$ respectively.

this[31] by projecting the probe beam to the position quadrature eigenstate $|X\rangle\langle X|$ with a measurement result X. The resulting two qubit state is then[31]

$$|\psi_X\rangle = [\beta_0|HH\rangle + \beta_3|VV\rangle]\langle X|\alpha_c\rangle_p + \beta_1|HV\rangle\langle X|\alpha_c e^{i\theta}\rangle_p + \beta_2|VH\rangle\langle X|\alpha_c e^{-i\theta}\rangle_p . \tag{63}$$

Now using the expression

$$\langle x|\alpha\rangle = \frac{1}{(2\pi)^{1/4}}\exp\left[-\mathrm{Im}(\alpha)^2 - \frac{(x-2\alpha)^2}{4}\right] \tag{64}$$

we obtain

$$|\psi_X\rangle = f(X,\alpha_c)[\beta_0|HH\rangle + \beta_3|VV\rangle] + f(X,\alpha_c\cos\theta)\left[\beta_1 e^{i\phi(X)}|HV\rangle + \beta_2 e^{-i\phi(X)}|VH\rangle\right] , \tag{65}$$

where, assuming that α_c is real,

$$f(x,\beta) = \exp\left[-\frac{1}{4}(x-2\beta)^2\right]/(2\pi)^{1/4} \tag{66}$$

$$\phi(x) = \alpha_c\sin\theta(x - 2\alpha_c\cos\theta) \bmod 2\pi . \tag{67}$$

Now $f(X,\alpha_c)$ and $f(X,\alpha_c\cos\theta)$ are two Gaussian curves with the mid point between the peaks located at $X_0 = \alpha_c[1+\cos\theta]$ and the peaks separated by a distance $X_d = 2\alpha_c[1-\cos\theta]$. As long as this difference is large $X_d \sim \alpha_c\theta^2 \gg 1$, then there is little overlap between these distributions and so our measurement can project the system into one of the two subspaces. The probability of a discrimination error occurring is given by

$$P_{\mathrm{error}} = \frac{1}{2}\mathrm{Erfc}[X_d/2\sqrt{2}] \tag{68}$$

which is less than 10^{-4} when $X_d > 2\pi$. For a measurement result X such that $X > X_0$, our solution (65) collapses to the even parity state

$$|\psi_{X>X_0}\rangle \sim \beta_0|HH\rangle + \beta_3|VV\rangle \tag{69}$$

while for $X < X_0$ we get the odd parity state

$$|\psi_{X<X_0}\rangle \sim \beta_1 e^{i\phi(X)}|HV\rangle + \beta_2 e^{-i\phi(X)}|VH\rangle . \tag{70}$$

The action of this two mode polarization non-demolition parity gate is clear. It splits the even parity terms (69) nearly deterministically from the odd parity cases (70). The odd parity state (70) depends on the value of the measured quadrature X. Simple local rotations using phase shifters that dependent on the measurement result X can be performed via a classical feedforward process to transform to

$$|\psi_{X<X_0}\rangle \sim \beta_1|HV\rangle + \beta_2|VH\rangle \tag{71}$$

which is independent of X. This correction operation requires that we are able to precisely determine X and so accurately determine the phase correction $\phi(X)$. The particular $\phi(X)$ is only known after the QND measurement and hence requires fast feedforward operations. This is not difficult but could limit the speed of operation of the gate.

There is another mechanism that eliminates the need for this phase correction but at the expense of two more weak nonlinearities. The scheme is depicted in Figure (17) and works as follows:

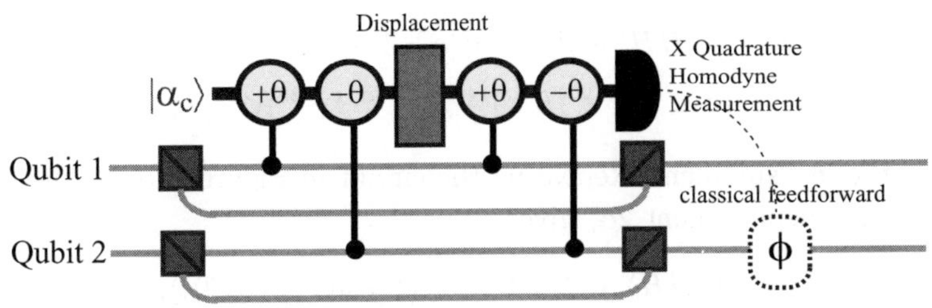

Fig. 17. Schematic diagram of a two-qubit polarization parity gate which does require the measurement-dependent phase correction.

Consider a two-qubit polarization-encoded state (61). After the interaction with the first two weak nonlinearities (plus a $-\theta$ phase rotation on the probe beam), our combined two qubit and probe system has the form

$$|\psi\rangle = [\beta_0|HH\rangle + \beta_3|VV\rangle]\,|\alpha_c\rangle_p + \beta_1|HV\rangle|\alpha_c e^{i\theta}\rangle_p + \beta_2|VH\rangle|\alpha_c e^{-i\theta}\rangle_p \quad (72)$$

We now displace the probe beam by an amount $-2\alpha_c\cos\theta$ and the system evolves to

$$\begin{aligned}|\psi\rangle &= [\beta_0|HH\rangle + \beta_3|VV\rangle]\,|\alpha_c\,(1-2\cos\theta)\rangle_p \\ &+ \beta_1 e^{i\alpha_c^2\sin 2\theta}|HV\rangle|\alpha_c\left(e^{i\theta}-2\cos\theta\right)\rangle_p \\ &+ \beta_2 e^{-i\alpha_c^2\sin 2\theta}|VH\rangle|\alpha_c\left(e^{-i\theta}-2\cos\theta\right)\rangle_p\,.\end{aligned} \quad (73)$$

Now applying the last two weak cross-Kerr nonlinearities and a phase shift $-\theta$ on the probe beam gives

$$\begin{aligned}|\psi\rangle &= [\beta_0|HH\rangle + \beta_3|VV\rangle]\,|\alpha_c\,(1-2\cos\theta)\rangle_p \\ &+ \left[\beta_1 e^{i\alpha_c^2\cos 2\theta}|HV\rangle + \beta_2 e^{-i\alpha_c^2\cos 2\theta}|VH\rangle\right]|-\alpha_c\rangle_p\,.\end{aligned} \quad (74)$$

We see immediately that our probe beam now has only two state components, $|-\alpha_c\rangle_p$ and $|\alpha_c(2\cos\theta-1)\rangle_p$, and hence an appropriate probe measurement will project us into the even or odd parity qubit subspace, without the need for a measurement-dependent phase correction. The cost is that two extra weak nonlinearities are needed. We do have a phase shift on the odd parity subspace that needs to be removed, but this can be done easily as it does not depend on the measurement outcome. It is also useful to note that the separation between the two coherent states scales as $\alpha_c\theta^2$ for $\theta \ll 1$. This can be improved with a simple trick, but at the expense of more nonlinearities. Let us consider the state (74) and displace the probe beam by $2\alpha_c$. This results in the state

$$\begin{aligned}|\psi\rangle &= [\beta_0|HH\rangle + \beta_3|VV\rangle]\,|\alpha_c\,(3 - 2\cos\theta)\rangle_p \\ &+ \left[\beta_1 e^{i\alpha_c^2\cos 2\theta}|HV\rangle + \beta_2 e^{-i\alpha_c^2\cos 2\theta}|VH\rangle\right]|\alpha_c\rangle_p \end{aligned} \tag{75}$$

before the measurement. Repeating the circuit in Figure 17 with the additional final displacement $2\alpha_c$ gives

$$\begin{aligned}|\psi\rangle &= [\beta_0|HH\rangle + \beta_3|VV\rangle]\,|\alpha_c\,(1 + 4(1 - \cos\theta))\rangle_p \\ &+ \left[\beta_1 e^{i\alpha_c^2\cos 2\theta}|HV\rangle + \beta_2 e^{-i\alpha_c^2\cos 2\theta}|VH\rangle\right]|\alpha_c\rangle_p\,. \end{aligned} \tag{76}$$

Repeating this circuit a total of n times gives

$$\begin{aligned}|\psi\rangle &= [\beta_0|HH\rangle + \beta_3|VV\rangle]\,|\alpha_c\,(1 + 2n(1 - \cos\theta))\rangle_p \\ &+ \left[\beta_1 e^{in\alpha_c^2\cos 2\theta}|HV\rangle + \beta_2 e^{-in\alpha_c^2\cos 2\theta}|VH\rangle\right]|\alpha_c\rangle_p \end{aligned} \tag{77}$$

and so we see our two coherent states are now separated by $2n\alpha_c\,(1 - \cos\theta) \sim n\alpha_c\theta^2$. We see that the use of the circuit n times allows us to increase the separation by a factor of n. The total number of nonlinearities used is now $4n$ but this has the advantage that α_c can be smaller.

So far all of our measurements on the probe beam have used standard homodyne measurements. Whilst these measurements are in principle straightforward to implement, they are by no means optimal. A near-optimal measurement is preferable as this enables the strength of the cross-Kerr nonlinearities used to be near-minimal[44]. Let us consider the combined state of the system in Figure 15 before the measurement and apply a displacement operation $D(-\alpha)$ on the probe beam. This results in the state

$$\begin{aligned}|\Psi\rangle_{abp} &= [\beta_0|HH\rangle_{ab} + \beta_3|VV\rangle_{ab}]\,|0\rangle_p \\ &+ e^{-i\alpha^2\sin\theta}\beta_1|HV\rangle_{ab}|\alpha_c(e^{i\theta} - 1)\rangle_p \\ &+ e^{i\alpha^2\sin\theta}\beta_2|VH\rangle_{ab}|\alpha_c(e^{-i\theta} - 1)\rangle_p\,. \end{aligned} \tag{78}$$

The $|HV\rangle_{ab}$ and $|VH\rangle_{ab}$ amplitudes have picked up a phase shift due to the displacement, which can be simply removed using linear optical elements. As long as $\alpha_c\theta \geq \pi$, a QND photon number measurement on the probe beam will project the a and b mode state into

$$|\Psi\rangle_{ab} = \beta_0|HH\rangle_{ab} + \beta_3|VV\rangle_{ab} \quad (79)$$

for a QND photon number measurement n_p of zero and

$$|\Psi\rangle_{ab} = \beta_1 e^{i\phi(n_p)}|HV\rangle_{ab} + \beta_2 e^{-i\phi(n_p)}|VH\rangle_{ab} \quad (80)$$

for $n_p > 0$ where $\phi(n_p) \sim -n_p\frac{\pi}{2}$. The error in discriminating the two components (even and odd parity states) is approximately $P_{err} \approx 10^{-4}$ for $\alpha\theta \sim \pi$. This is a near optimal measurement and can be realised if one has a weak nonlinearity and efficient homodyne measurements[44].

For all three approaches above we have chosen to call the even parity state $\{|HH\rangle, |VV\rangle\}$ and the odd parity states $\{|HV\rangle, |VH\rangle\}$, but this is an arbitrary choice, primarily dependent on the form or type of PBS used to convert the polarization encoded qubits to which path encoded qubits. Any other choice is also acceptable and it does not have to be the same between the two qubits.

5. Bell state measurements

These near-deterministic non-destructive parity measurements are critically important in optical quantum information processing. They can be used to build a number of interesting devices and gates. For instance, with a non-destructive parity gate it is possible to build a Bell-state detector. Bell-state measurements are known to be one of the essential and enabling tools in quantum computation and communication. If we consider the four Bell states

$$|\Psi^+\rangle \equiv \frac{1}{\sqrt{2}}\left(|H,V\rangle + |V,H\rangle\right) \quad (81)$$

$$|\Psi^-\rangle \equiv \frac{1}{\sqrt{2}}\left(|H,V\rangle - |V,H\rangle\right) \quad (82)$$

$$|\Phi^+\rangle \equiv \frac{1}{\sqrt{2}}\left(|H,H\rangle + |V,V\rangle\right) \quad (83)$$

$$|\Phi^-\rangle \equiv \frac{1}{\sqrt{2}}\left(|H,H\rangle - |V,V\rangle\right) \quad (84)$$

then it is clear that our standard parity gate will distinguish $|\Phi^\pm\rangle$ from $|\Psi^\pm\rangle$. Hence an application of the parity detector distinguishes two subclasses of the Bell states.

The natural question now becomes: can we modify the parity gate to distinguish $|\Psi^+\rangle$ from $|\Psi^-\rangle$ and $|\Phi^+\rangle$ from $|\Phi^-\rangle$? In the parity gate used above the $\{H, V\}$ polarizing beam-splitters dictates that $\{|H,H\rangle, |V,V\rangle\}$ are distinguished from $\{|H,V\rangle, |V,H\rangle\}$. Bell states can be written in the $\{D = H + V, \bar{D} = H - V\}$ basis as

$$|H,V\rangle + |V,H\rangle \to |D,D\rangle - |\bar{D},\bar{D}\rangle \tag{85}$$

$$|H,V\rangle - |V,H\rangle \to |D,\bar{D}\rangle - |\bar{D},D\rangle \tag{86}$$

$$|H,H\rangle + |V,V\rangle \to |D,D\rangle + |\bar{D},\bar{D}\rangle \tag{87}$$

$$|H,H\rangle - |V,V\rangle \to |D,\bar{D}\rangle + |\bar{D},D\rangle \tag{88}$$

With respect to $\{D, \bar{D}\}$, $|\Psi^+\rangle$ and $|\Phi^+\rangle$ are even parity and $|\Psi^-\rangle$ and $|\Phi^-\rangle$ are odd parity. Hence the $\{D, \bar{D}\}$ parity gate will allow us to distinguish $\{|\Phi^+\rangle, |\Psi^+\rangle\}$ from $\{|\Phi^-\rangle, |\Psi^-\rangle\}$.

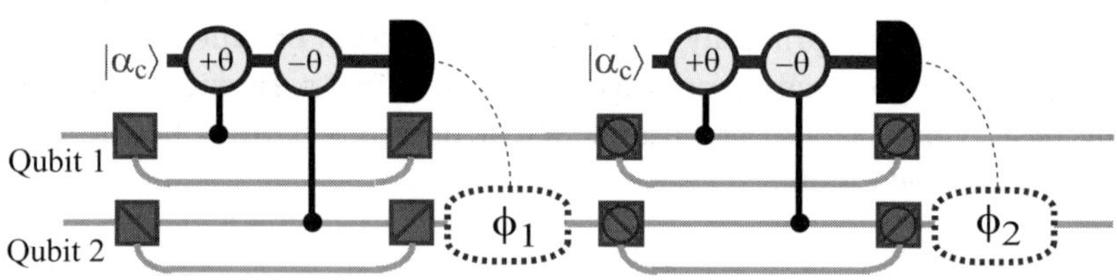

Fig. 18. Schematic diagram of a non-destructive Bell state measurement, composed of two QND parity detectors. The first parity gate uses the standard $\{H, V\}$ PBS and distinguishes the $|\Phi^{\pm}\rangle$ Bell states from the $|\Psi^{\pm}\rangle$ ones. An even-parity result for gate indicates the presence of $|\Phi^{\pm}\rangle$ while an odd-parity result indicates the presence of $|\Psi^{\pm}\rangle$. For the latter result a local operation on the second qubit is required to remove the phase $\phi(X_1)$ induced by the measurement. Once this correction is done the second parity gate can be applied. This gate is similar to the first one but has 45-deg PBS's (square box with circle inside) that operate in the $\{D, \bar{D}\}$ basis. An even-parity result indicates the presence of $|\Phi^+\rangle$ or $|\Psi^+\rangle$ while an odd parity result indicates the presence of the $|\Phi^-\rangle$ or $|\Psi^-\rangle$. Again a phase correction $\phi(X_2)$ in the $\{D, \bar{D}\}$ basis is needed for an odd-parity result, to remove the unwanted phase shift.

Now since the both ($\{H, V\}$ and $\{D, \bar{D}\}$) parity gates are nondestructive on the qubits and select different pairs of Bell states, they allow a natural construction of a Bell-state analyser (depicted in Figure 18). From each parity measurement we get one bit of information, indicating whether the parity was even or odd, and so from both parity measurements we end up with four possible results (even, even), (even, odd), (odd, even) and (odd, odd). This is enough to uniquely identify all the Bell states. It is important to remove the unwanted phase factors that have arisen after an odd parity measurement result. This needs to be done in the same basis as the PBS in the particular parity gate. For instance, for an odd parity result giving $X = X_1$ on the first parity gate, a phase shift $\phi(X_1)$ needs to be removed in the $\{H, V\}$ basis. Similarly for a odd parity results giving $X = X_2$ on the second parity gate a phase shift $\phi(X_2)$ needs to be removed in the $\{D, \bar{D}\}$ basis.

So far we have shown how it is possible, using linear elements, weak cross-Kerr nonlinearities and homodyne measurements, to create a wide range of high efficiency quantum detectors and gates that can perform tasks ranging from photon-number discrimination to Bell-state measurements. This is all achieved non-destructively on the photonic qubits and so provides a critical set of tools, extremely useful for single-photon quantum computation and communication. With these tools, universal quantum computation can be achieved using the ideas and techniques originally proposed by KLM.

6. A resource-efficient CNOT gate

The parity gate and the Bell-state analyser have shown the versatility of an approach using weak nonlinearities and homodyne conditioning measurements. Both of these gates/detectors can be used to induce two-qubit operations and hence are all that is necessary, with single-qubit operation and single-photon measurements, to perform universal quantum computation. However, the parity and Bell-state gates are not the typical two-qubit gates that one generally considers in the standard quantum computational models. The typical two-qubit gate generally considered is the CNOT gate. This can be constructed from two parity gates (like the Bell-state detector) but requires an ancilla qubit. This CNOT gate is depicted schematically in Figure 19) and operates as follows:

Suppose that the control and target qubits are initially in the joint state

$$c_0|HH\rangle_{ct} + c_1|HV\rangle_{ct} + c_2|VH\rangle_{ct} + c_3|VV\rangle_{ct} \tag{89}$$

with the ancilla mode prepared in $|H\rangle_a + |V\rangle_a$. The action of the first parity gate between the control and ancilla mode (c and a) conditions the system to

$$c_0|HHH\rangle_{cat} + c_1|HHV\rangle_{cat} + c_2|VVH\rangle_{cat} + c_3|VVV\rangle_{cat} \tag{90}$$

for an even-parity result and

$$c_0 e^{i\phi(X_1)}|HVH\rangle_{cat} + c_1 e^{i\phi(X_1)}|HVV\rangle_{cat} + c_2 e^{-i\phi(X_1)}|VHH\rangle_{cat} + c_3 e^{-i\phi(X_1)}|VHV\rangle_{cat} \tag{91}$$

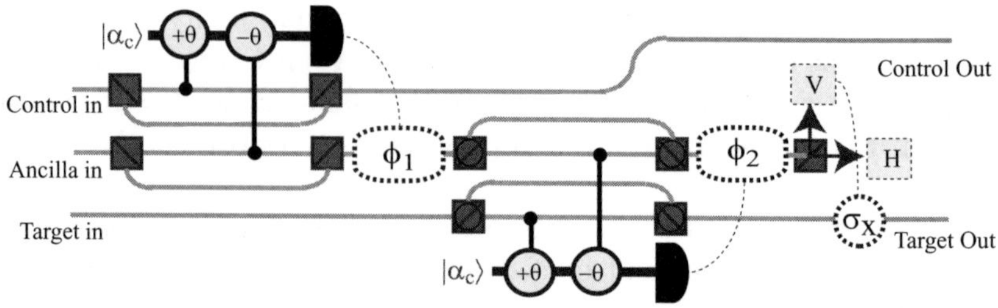

Fig. 19. Schematic diagram of a near-deterministic CNOT composed of two parity gates (one with PBS in the $\{H, V\}$ basis and one with PBS in the $\{D, \bar{D}\}$ basis), one ancilla qubit prepared initially in $|H\rangle + |V\rangle$, a polarization-determining photon-number QND measurement and classical feedforward elements.

for an odd-parity result. The unwanted phase $\phi(X_1)$ can be removed easily using linear optics, and then a bit flip on the ancilla state leads to (90). Now let us look at the action of the second parity gate, which operates with $\{D, \bar{D}\}$ basis polarizing beam-splitters on the ancilla and target qubits. It is useful to rewrite the ancilla and target modes of (90) in the basis of $\{D, \bar{D}\}$ as

$$\begin{aligned} &[(c_0 + c_1)\,|H\rangle_c + (c_2 + c_3)\,|V\rangle_c]\,|DD\rangle_{at} \\ &+ [(c_0 - c_1)\,|H\rangle_c - (c_2 - c_3)\,|V\rangle_c]\,|\bar{D}\bar{D}\rangle_{at} \\ &+ [(c_0 - c_1)\,|H\rangle_c + (c_2 - c_3)\,|V\rangle_c]\,|D\bar{D}\rangle_{at} \\ &+ [(c_0 + c_1)\,|H\rangle_c - (c_2 + c_3)\,|V\rangle_c]\,|\bar{D}D\rangle_{at}\,. \end{aligned} \tag{92}$$

The action of the $\{D, \bar{D}\}$ parity gate is now clear. For an even-parity result the above state is projected to

$$\begin{aligned} &[(c_0 + c_1)\,|H\rangle_c + (c_2 + c_3)\,|V\rangle_c]\,|DD\rangle_{at} \\ &+ [(c_0 - c_1)\,|H\rangle_c - (c_2 - c_3)\,|V\rangle_c]\,|\bar{D}\bar{D}\rangle_{at} \end{aligned} \tag{93}$$

while for an odd-parity result (after phase correction) we obtain

$$\begin{aligned} &[(c_0 - c_1)\,|H\rangle_c + (c_2 - c_3)\,|V\rangle_c]\,|D\bar{D}\rangle_{at} \\ &+ [(c_0 + c_1)\,|H\rangle_c - (c_2 + c_3)\,|V\rangle_c]\,|\bar{D}D\rangle_{at}\,. \end{aligned} \tag{94}$$

Now this odd parity state can be transformed to the even parity case by bit-flipping the ancilla qubit and performing a sign flip on the $|V\rangle$ component of the control. After such an operation, the state is given by (93). Now performing a measurement of the ancilla mode in the $\{H, V\}$ basis, the control and target qubits are transformed to

$$c_0|HH\rangle_{ct} + c_1|HV\rangle_{ct} + c_2|VV\rangle_{ct} + c_3|VH\rangle_{ct} \tag{95}$$

for an H result, and

$$c_0|HV\rangle_{ct} + c_1|HH\rangle_{ct} + c_2|VH\rangle_{ct} + c_3|VV\rangle_{ct} \tag{96}$$

for a V result. The second case can be transformed to the first by a bit flip of the target qubit. After all such operations and phase corrections, our initial control and target qubits have been transformed as

$$\begin{aligned} c_0|HH\rangle_{ct} + c_1|HV\rangle_{ct} &+ c_2|VH\rangle_{ct} + c_3|VV\rangle_{ct} \\ &\rightarrow c_0|HH\rangle_{ct} + c_1|HV\rangle_{ct} + c_2|VV\rangle_{ct} + c_3|VH\rangle_{ct} \,. \end{aligned} \tag{97}$$

This is clearly the result one would expect if a CNOT operation had been performed on the control and target qubits and shows how our weak nonlinearity approach can implement a near deterministic CNOT operation, utilizing only one ancilla qubit (which is not destroyed at the end of the gate and could in principle be re-used). This represents a huge saving in the physical resources to implement single-photon quantum logic gates, compared to the previous linear optics schemes.

6.1. *A discussion on the weak nonlinearity approach*

We have shown how it is possible to create near deterministic two-qubit gates (parity, Bell and CNOT) without a huge overhead in ancilla resources. In fact, an ancilla photon is required only for the CNOT gate. The key additions to the general linear optical resources are weak cross-Kerr nonlinearities and efficient homodyne measurements. Homodyne measurements are a well established technique, frequently used in the continuous variable quantum information processing community. However weak cross-Kerr nonlinearities are not commonly used within optical quantum computational devices and as such a discussion of the source and strength of such elements is required. Before this we really need to define what we mean by *weak*—weak compared with what? Basically it is well known that deterministic two-qubit gates can be performed if one has access to a cross-Kerr nonlinearity that can induce a π phase shift directly between single photons. This leads to a natural definition of weak nonlinearities, that is, the use of nonlinear cross-Kerr materials (when all are taken into account) that cannot directly induce a phase shift of the order of π. This seems to give an acceptable functional definition.

For the majority of the parity gates discussed previously we have established that the nonlinearity θ must satisfy the constraint $\alpha_c\theta^2 \sim 8$ where α_c is the amplitude of the probe beam. For a weak nonlinearity $\theta \ll 1$ we must choose $\alpha_c \sim 10/\theta^2$, so for instance if $\theta \sim 10^{-2}$ then $\alpha_c \geq 10^5$

(which corresponds to a probe beam with mean photon number 10^{10}). For a smaller θ we need a much larger α_c. This puts a natural constraint on θ, since α_c cannot be made arbitrarily large in practice. For the most efficient parity-based gates discussed previously we have established that the nonlinearity θ must satisfy the constraint $\alpha\theta \sim \pi$ where, just to re-emphasize, α is the amplitude of the probe beam. Thus, due to the weak nature of the nonlinearity $\theta \ll 1$ we must choose $\alpha \sim \pi/\theta$, so for instance if $\theta \sim 10^{-5}$ then $\alpha \geq \pi 10^5$ (which corresponds to a probe beam with mean photon number 10^{11}).

This leads to the question of a mechanism to achieve the weak cross-Kerr nonlinearity. Natural χ^3 materials have small nonlinearities[32] on the order of 10^{-18} which would require lasers with $\alpha_c \sim 10^{37}$, which is physically unrealistic. However, systems such as optical fibers[33], silica whispering-gallery micro-resonators [34], cavity QED systems[35,36] and EIT[37] are capable of producing much larger nonlinearities. For example, calculations for EIT systems in NV diamond[30] have shown that potential phase shifts of order $\theta = 0.01$ are achievable. With $\theta = 0.01$ the probe beam must have an amplitude of at least 10^3, which is physically reasonable with current technology.

Finally, by using these weak cross-Kerr nonlinearities to aid the construction of near deterministic two-qubit gates, we can build quantum circuits with far fewer resources than are required for the current corresponding approaches with linear optics. This has enormous implications for the development of single-photon quantum computing and information processing, using either the conventional gate-based models, or cluster-state techniques. The approach can be applied directly to optical cluster-state computation, allowing a significant reduction in the physical resources needed. It is straightforward to show that in principle an n-qubit computation only requires of order n single photon sources. This truly indicates the power of a weak nonlinearity approach. The strength of the nonlinearities required for our gate are also orders of magnitude weaker than those required to perform CNOT gates directly between the single photons.

7. Concluding discussion: is optical computation possible?

We hope that so far we have shown the progress in using optical systems for quantum computation (QC) and quantum information progressing (QIP). Optical QIP is currently a very active research area, both theoretically and experimentally. To be able to perform universal quantum computation with optical fields (either at the single photon level or continuous variable level) it is known that nonlinearities are needed. The form of the required optical nonlinearities doesn't really matter; for example they can be generated via measurement. The work of Knill, Laflamme and Milburn (KLM) has shown that in principle universal quantum computation is possible with linear optics [17] and using such nonlinearities. However, due to the probabilistic nature of the gates in linear optical QIP, it is practically rather inefficient (in terms of photon resources) to implement[39–41]. Strong cross-Kerr nonlinearities are able to effectively mediate an interaction directly between photonic qubits. This would realize deterministic quantum gates and thus efficient optical QIP. However, in practice, such nonlinearities are not available. On the other hand, much weaker nonlinearities can be generated, for example, with electromagnetically induced transparency (EIT)[37,38,30]. We have shown how such weak nonlinearities provide the building blocks for efficient optical QIP[35,31,42–46].

There has also been a significant amount of experimental progress in demonstrating optical gates for quantum computation. The last five years has seen a number of key demonstrations, including:

- A destructive two-photon parity and CNOT gate[47,48]
- A three-photon destructive CNOT gate[49]
- The nonlinear sign-shift gate[50]
- The Franson four-photon CNOT gate[51,52]
- A complete Bell-state measurement[53]
- Teleportation and entanglement swapping with polarization-encoded qubits[54,55]
- Generation of the three- and four-photon cluster states[56,57]
- Encoding qubits for error prevention[58,59]
- Continuous-variable teleportation and entanglement swapping[60–63]
- The generation of weak nonlinearities and the slowing of light[36,64]

All of these demonstrations show the potential of optics for information processing. What we have not discussed so far is the generation of relevant initial quantum states and their detection. To do this we need to consider the single-photon and continuous-variable situations separately. We have discussed the single-photon detection requirement previously, but not how this can be achieved.

In the single-photon qubit regime we need two critical resources to enable scalable quantum computation: a generator of single photons on demand and a detector to measure whether a photon is present or not. With these resources and linear optical elements, it is straightforward to create an arbitrary polarization-encoded qubit and to detect it. The key is the generation of the on-demand source and the photon detection. Currently both of these resources are still in development[65] but are not efficient enough yet. However, the progress is very promising.

In the continuous variable regime the generation of an initial state and the measurement of the qunat in the computational basis are rather straightforward. If we encode our qunat in terms of the X and P quadratures of the electromagnetic field, then for the initialization we need to be able to generate a position or momentum quadrature eigenstate. The vacuum state $|0\rangle$ is ideal for this, and simply generated. The measurement of the qunat in the computational basis can be achieved via a highly efficient homodyne measurement (depicted schematically in Figure 20). The homodyne measurement works as follows: Consider a signal mode specified by the creation and destruction operators $a^\dagger$ and a and an intense local oscillator $\epsilon e^{i\theta}$ where θ is the local oscillator phase.. If the signal mode and oscillator are mixed on a 50/50 beam-splitter and then the output mode intensities are measured on photodiodes, it is straightforward to show that the difference in the currents from the detectors C and D is

$$\begin{aligned} I_d &= \langle c^\dagger c\rangle - \langle d^\dagger d\rangle \\ &= \eta\epsilon\left(a^\dagger e^{-i\theta} + a e^{i\theta}\right) \\ &= \eta\epsilon X\theta \end{aligned} \tag{98}$$

where $X_0 = X$ and $X_{\pi/2} = P$, and η is the efficiency of the detectors. Thus as long as one knows η and ϵ, a measurement of the current difference measures X or P depending on the local oscillator phase. The homodyne measurements currently have overall efficiencies exceeding 99%[66].

Now to complete our discussion of optics for quantum computation let us summarize how well the DiVincenzo criteria are satisfied, for both our single-photon and continuous-variable variable computation schemes.

7.1. *The DiVincenzo criteria for single photons*

For single-photon quantum computation the status of the DiVincenzo criteria is as follows:

- We have well defined qubits in terms of the polarization degree of freedom of single photons, which is equivalent to which-path encoding.

- The initialization of qubits corresponds to creating single-photon states with well-defined polarizations. Such single-photon sources are cur-

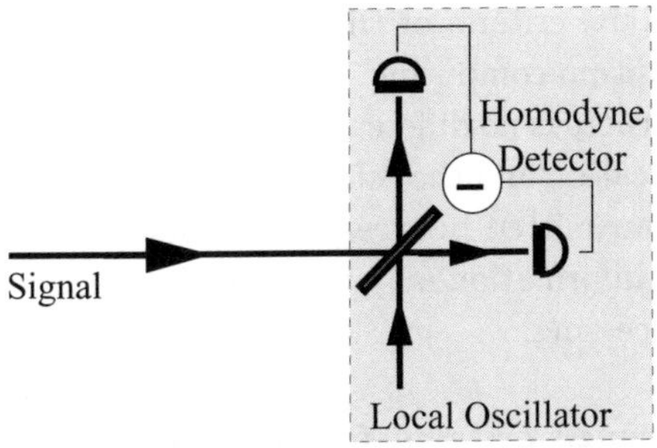

Fig. 20. Schematic diagram of a homodyne measurement. The homodyne detector is composed of two optical modes: a signal mode and an intense local oscillator. Both these modes are combined on a 50/50 beam-splitter and the difference in photocurrents between the detectors on the output modes is measured.

rently under active development and are promising: At this moment, there exist sources that have sufficient coherence for self-interference.

- Photons have extremely long coherence times, and suffer from low loss when traveling through optical fibers or free space.

- For universal computation we need photons to interact with each other. A two-qubit interaction is essential. In free space photons do not interact with each other; however, we can induce a probabilistic two-photon gate by using the bunching effect at beam-splitters and projective measurements, or via nonlinear effects (such as those generated by EIT). These nonlinear effects do not necessarily need to be strong; weak nonlinearities with a coherent field acting as a bus can be used to implement near deterministic CNOT gates between photonic qubits. Alternatively the one-way or cluster-state quantum computational models can be used.

- Next, we need the ability to faithfully read the quantum state of a qubit in the computational basis. In the single-photon case this is achieved through photodetection. However, current detectors have less than ideal properties. Typical quantum efficiencies reach at most 90 percent, while we need at least 99 percent. There are, however, exciting new proposals for high-fidelity and high efficiency photodetectors, which don't even absorb the photons.

The two additional criteria of DiVincenzo are easily satisfied in single-photon optical quantum computing. Since single photons can be considered as "flying qubits", there is no difference between the flying qubits needed for quantum communication and the "stationary qubits" used for computation. Photons can be transmitted between remote locations. Static memory for photonic quantum information may be a useful asset in the future, for large scale quantum processing.

7.2. *The DiVincenzo criteria for continuous variables*

For continuous variables we may summarize the status of the DiVincenzo criteria as follows:

- We need to relax the criterion for qubits. We require a collection of well-characterized qu-units (qu-bit, qu-dit, and qunat). For continuous-variables (qunats) we can use X and P (the quadratures) of the electromagnetic fields to encode our quantum information.

- State preparation is straightforward as the vacuum of the electromagnetic field can be used. No preparation is required to create the vacuum.

- Optical systems tend to have long coherence times. However the loss of a single photon in CV systems can be problematic. The exact coherence properties depend on the physical implementation, that is whether it is in free space or in a fiber.

- We need to define what we mean by quantum computation, as a continuous variable needs an infinite number of parameters to describe it. We demand that we can create Hamiltonians that are polynomial functions of the operators corresponding to the continuous variables. Such Hamiltonians should be of degree higher than quadratic (if homodyne measurements are used), otherwise the system could be efficiently classically simulated.

- The choice of measurement is homodyne/heterodyne measurements. These are projective measurements in the X (and/or P) basis and have been implemented with very high efficiency. For optical frequencies the efficiencies exceed 99 percent at present[66].

So to finally conclude, we believe that optical systems are ideally suited for quantum communication and are a very promising candidate for quantum computation. It is very likely that fully quantum information processing devices will contain optical components at their core.

8. Acknowledgments

We thank S. D. Barrett, R. G. Beausoleil, P. Kok and G. J. Milburn for numerous valuable discussions. This work was supported in part by Japanese JSPS, MPHPT, and Asahi-Glass research grants and the European Project RAMBOQ.

References

1. M. A. Nielsen and I. L. Chuang, *Quantum Computation and Quantum Information*, (Cambridge University Press, Cambridge, U.K., 2000); H.-K. Lo, S. Popescu and T. P. Spiller (eds.), *Introduction to Quantum Computation and Information*, (World Scientific Publishing, 1998).
2. N. Gisin, G. Ribordy, W. Tittel, and H. Zbinden, Rev. Mod. Phys. **74**, 145 (2002).
3. R. G. Clark (ed.), *Experimental Implementation of Quantum Computation*, (Rinton Press, 2001), ISBN 1-58949-013-4.
4. N. A. Gershenfeld and I. L. Chuang, Science **275**, 350 (1997); D. G. Cory, A. F. Fahmy, and T. F. Havel, Proc. Nat. Acad. Sci. **94**, 1634 (1997); Y. Nakamura, Yu. A. Pashkin and J. S. Tsai, Nature **398**, 786 (1999); D. Vion, A. Aassime, A. Cottet, P. Joyez, H. Pothier, C. Urbina, D. Esteve, and M. H. Devoret, Science **296**, 886 (2002); Y. Yu, S. Han, X. Chu, S. Chu and Z. Wang, Science **296**, 889 (2002); B. E. Kane, Nature **393**, 133 (1998).
5. See for instance: P. G. Kwiat, K. Mattle, H. Weinfurter, A. Zeilinger, A. V. Sergienko and Y. Shih, Phys. Rev. Lett. **75**, 4337 (1995); W. Tittel, J. Brendel, H. Zbinden and N. Gisin, Phys. Rev. Lett. **81**, 3563 (1998); P. G. Kwiat, E. Waks, A. G. White, I. Appelbaum and P. H. Eberhard, Phys. Rev. A **60**, R773 (1999); G. Weihs, T. Jennewein, C. Simon, H. Weinfurter and A. Zeilinger, Phys. Rev. Lett. **81**, 5039 (1998).
6. D. P. DiVincenzo, "Topics in Quantum Computers", in Mesoscopic Electron Transport, (ed. Kowenhoven, L., Schn, G. & Sohn, L.), NATO ASI Series E, (Kluwer Ac. Publ., Dordrecht, 1997); cond-mat/9612126.
7. D. P. DiVincenzo, Fortschr. Phys. **48**, 9 (2000).
8. P. W. Shor, Phys. Rev. A **52**, 2493 (1995); P. W. Shor, "Fault-tolerant quantum computation", Proc. 37th Annual Symposium on the Foundations of Computer Science, 56 (IEEE Computer Society Press, Los Alamitos, CA, 1996), quant-ph/9605011; A. M. Steane, Phys. Rev. Lett. **77**, 793 (1996); A. M. Steane, Phys. Rev. Lett. **78**, 2252 (1997).
9. D. Gottesman, in *Proceedings of the XXII International Colloquium on Group Theoretical Methods in Physics*, edited by S. P. Corney et al. (International Press, Cambridge, MA, 1999), p. 32.
10. M. A. Nielsen and I. L. Chuang, *Quantum Computation and Quantum Information*, (Cambridge University Press, Cambridge, U.K., 2000), p 464.

11. S. D. Bartlett, B, C. Sanders, S. L. Braunstein and Kae Nemoto, Phys. Rev. Lett. **88**, 097904 (2002).
12. T. Rudolph and L. Grover, *A 2 rebit gate universal for quantum computing*, quant-ph/0210187.
13. M. A. Nielsen, Phys. Lett. A **308**, 96 (2003).
14. D. W. Leung, Int. J. Quant. Inf. **2**, 33 (2004).
15. R. Raussendorf and H. J. Briegel, Phys. Rev. Lett. **86**, 5188 (2001).
16. Kae Nemoto and W. J. Munro, *Universal quantum computation on the power of quantum non-demolition measurements*, accepted to Phys. Lett A (2005).
17. E. Knill, R. Laflamme and G. Milburn, Nature **409**, 46 (2001).
18. T. C. Ralph, A. G. White, W. J. Munro and G. J. Milburn, Phys. Rev. A **65**, 012314 (2002).
19. T. Rudolph and Jian-Wei Pan, *A simple gate for linear optics quantum computing*, quant-ph/0108056.
20. E. Knill, Phys. Rev. A **66**, 052306 (2002).
21. T.B. Pittman, B.C. Jacobs and J.D. Franson, Phys. Rev. A **64**, 062311 (2001).
22. D. Gottesman and I. L. Chuang, Nature **402**, 390 (1999).
23. C. H. Bennett, G. Brassard, C. Crépeau, R. Jozsa, A. Peres, and W. K. Wootters , Phys. Rev. Lett. **70**, 1895 (1993).
24. M. Nielsen, Phys. Rev. Lett. **93**, 040503 (2004).
25. D. E. Browne and T. Rudolph, Phys. Rev. Lett. **95**, 010501 (2005).
26. R. G. Beausoleil, W. J. Munro and T. P. Spiller, J. Mod. Opt. **51**, 1559 (2004).
27. S. E. Harris and Y. Yamamoto, Phys. Rev. Lett. **81**, 3611 (1998).
28. A. Gilchrist, G. J. Milburn, W. J. Munro and Kae Nemoto, *Generating optical nonlinearity using trapped atoms*, quant-ph/0305167.
29. S. Lloyd and S. L. Braunstein, Phys. Rev. Lett. **82**, 1784 (1999).
30. W. J. Munro, K. Nemoto, R. G. Beausoleil and T. P. Spiller, Phys. Rev. A **71**, 033819 (2005).
31. S. D. Barrett, P. Kok, Kae Nemoto, R. G. Beausoleil, W. J. Munro and T. P. Spiller, Phys. Rev. A **71**, 060302 (2005).
32. P. Kok, H Lee and J. P. Dowling, Phys. Rev. A **66**, 063814 (2002).
33. X. Li, P. L. Voss, J. E. Sharping and P. Kumar, *Raman-noise induced quantum limits for* χ^3 *nondegenerate phase-sensitive amplification and quadrature squeezing*, quant-ph/0402191.
34. T. J. Kippenberg, S. M. Spillane and K. J. Vahala, Phys. Rev. Lett. **93**, 083904 (2004).
35. P. Grangier, J. A. Levenson and J.-P. Poizat, Nature **396**, 537 (1998).
36. Q. A. Turchette, C. J. Hood, W. Lange, H. Mabuchi and H. J. Kimble, Phys. Rev. Lett. **75**, 4710 (1995).
37. H. Schmidt and A. Imamoglu, Optics Letters **21**, 1936 (1996).
38. M. Paternostro, M. S. Kim and B. S. Ham, Phys. Rev. A **67**, 023811 (2002)).
39. S. Scheel, Kae Nemoto, W. J. Munro and P. L. Knight, Phys. Rev. A **68**, 032310 (2003).
40. S. Scheel, *Scaling of success probabilities for linear optics gates*, quant-

ph/0410014.
41. J. Eisert, *Optimizing linear optics quantum gates*, quant-ph/0409156.
42. Kae Nemoto and W. J. Munro, Phys. Rev. Lett **93**, 250502 (2004).
43. W. J. Munro, Kae Nemoto, T. P. Spiller, S. D. Barrett, P. Kok and R. G. Beausoleil, J. Opt. B: Quantum Semiclass. Opt. **7**, S135 (2005).
44. W. J. Munro, K. Nemoto and T. P. Spiller, New J. Phys. **7**, 137 (2005).
45. M. G. A. Paris, M. Plenio, D. Jonathan, S. Bose and G. M. D'Ariano, Phys Lett A **273**, 153 (2000).
46. G. M. D'Ariano, L. Maccone, M. G. A. Paris and M. F. Sacchi, Phys. Rev. A **61**. 053817 (2000).
47. T. B. Pittman, B. C. Jacobs and J. D. Franson, Phys. Rev. Lett. **88**, 257902 (2002).
48. J. L. O'Brien, G. J. Pryde, A. G. White, T. C. Ralph and D. Branning, Nature **426**, 264 (2003).
49. T. B. Pittman, M. J. Fitch, B. C. Jacobs and J. D. Franson, Phys. Rev. A **68**, 032316 (2003).
50. K. Sanaka, T. Jennewein, Jian-Wei Pan, K. Resch and A. Zeilinger, Phys. Rev. Lett. **92**, 017902 (2004).
51. S. Gasparoni, J. Pan, P. Walther, T. Rudolph and A. Zeilinger, Phys. Rev. Lett. **93**, 020504 (2004).
52. Zhi Zhao, An-Ning Zhang, Yu-Ao Chen, Han Zhang, Jiangfeng Du, Tao Yang and Jian-Wei Pan, Phys. Rev. Lett. **94**, 030501 (2005).
53. P. Walther and A. Zeilinger, Phys. Rev. A **72** 010302(R) (2005).
54. D. Bouwmeester, Jian-Wei Pan, K. Mattle, M. Eibl, H. Weinfurter and A. Zeilinger, Nature **390**, 575 (1997).
55. Jian-Wei Pan, S. Gasparoni, M. Aspelmeyer, T. Jennewein and A. Zeilinger, Nature **421**,721 (2003).
56. P. Walther, K. J.Resch, T. Rudolph, E. Schenck, H. Weinfurter, V. Vedral, M. Aspelmeyer and A. Zeilinger, Nature **434**, 169 (2005).
57. An-Ning Zhang, Chao-Yang Lu, Xiao-Qi Zhou, Yu-Ao Chen, Zhi Zhao, Tao Yang and Jian-Wei Pan, *Experimental Construction of Optical Multi-qubit Cluster States From Bell States*, quant-ph/0501036.
58. T. B. Pittman, B. C Jacobs and J. D. Franson, *Demonstration of Quantum Error Correction using Linear Optics*, quant-ph/0502042.
59. J. L. O'Brien, G. J. Pryde, A. G. White and T. C. Ralph, Phys. Rev. A **71**, 060303(R) (2005).
60. A. Furusawa, J. L. Srensen, S. L. Braunstein, C. A Fuchs, H. J. Kimble and E. S Polzik, Science **282**, 706 (1998).
61. H. Yonezawa, T. Aoki and A. Furusawa, Nature **431**, 430 (2004).
62. A. Dolinska, B. C. Buchler, P. K. Lam, T. C. Ralph and W. P. Bowen, Phys. Rev. A **68**, 052308 (2003).
63. N. Takei, H. Yonezawa, T. Aoki and A. Furusawa, Phys. Rev. Lett. **94**, 220502 (2005).
64. S. E. Harris and L. V. Hau, Phys. Rev. Lett. **82**, 4611 (1999); H. Altug and J. Vucdkovic, Appl. Phys. Lett. **86**, 111102 2005.
65. C. Santori, D. Fattal, J. Vuckovic, G.S. Solomon and Y. Yamamoto, Na-

ture **419**, 594 (2002); E.Waks, K. Inoue, W. D. Oliver, E.Diamanti and Y.Yamamoto, IEEE Journal of Selected Topics in Quantum Electronics **9**, 1502 (2003).

66. E. S. Polzik, J. Carry and H. J. Kimble, Phys. Rev. Lett. **68**, 3020 (1992).